Lena Uetzmann

Differentiation of the Endoderm Lineage in the Murine System

Lena Uetzmann

Differentiation of the Endoderm Lineage in the Murine System

in vivo and in vitro

Südwestdeutscher Verlag für Hochschulschriften

Impressum/Imprint (nur für Deutschland/ only for Germany)
Bibliografische Information der Deutschen Nationalbibliothek: Die Deutsche Nationalbibliothek verzeichnet diese Publikation in der Deutschen Nationalbibliografie; detaillierte bibliografische Daten sind im Internet über http://dnb.d-nb.de abrufbar.

Alle in diesem Buch genannten Marken und Produktnamen unterliegen warenzeichen-, marken- oder patentrechtlichem Schutz bzw. sind Warenzeichen oder eingetragene Warenzeichen der jeweiligen Inhaber. Die Wiedergabe von Marken, Produktnamen, Gebrauchsnamen, Handelsnamen, Warenbezeichnungen u.s.w. in diesem Werk berechtigt auch ohne besondere Kennzeichnung nicht zu der Annahme, dass solche Namen im Sinne der Warenzeichen- und Markenschutzgesetzgebung als frei zu betrachten wären und daher von jedermann benutzt werden dürften.

Verlag: Südwestdeutscher Verlag für Hochschulschriften Aktiengesellschaft & Co. KG
Dudweiler Landstr. 99, 66123 Saarbrücken, Deutschland
Telefon +49 681 37 20 271-1, Telefax +49 681 37 20 271-0
Email: info@svh-verlag.de
Zugl.: München, Technische Universität, Dissertation, 2009

Herstellung in Deutschland:
Schaltungsdienst Lange o.H.G., Berlin
Books on Demand GmbH, Norderstedt
Reha GmbH, Saarbrücken
Amazon Distribution GmbH, Leipzig
ISBN: 978-3-8381-1263-3

Imprint (only for USA, GB)
Bibliographic information published by the Deutsche Nationalbibliothek: The Deutsche Nationalbibliothek lists this publication in the Deutsche Nationalbibliografie; detailed bibliographic data are available in the Internet at http://dnb.d-nb.de.

Any brand names and product names mentioned in this book are subject to trademark, brand or patent protection and are trademarks or registered trademarks of their respective holders. The use of brand names, product names, common names, trade names, product descriptions etc. even without a particular marking in this works is in no way to be construed to mean that such names may be regarded as unrestricted in respect of trademark and brand protection legislation and could thus be used by anyone.

Publisher: Südwestdeutscher Verlag für Hochschulschriften Aktiengesellschaft & Co. KG
Dudweiler Landstr. 99, 66123 Saarbrücken, Germany
Phone +49 681 37 20 271-1, Fax +49 681 37 20 271-0
Email: info@svh-verlag.de

Printed in the U.S.A.
Printed in the U.K. by (see last page)
ISBN: 978-3-8381-1263-3

Copyright © 2010 by the author and Südwestdeutscher Verlag für Hochschulschriften Aktiengesellschaft & Co. KG and licensors
All rights reserved. Saarbrücken 2010

*Achte auf Deine Gedanken,
denn sie werden Worte.*

*Achte auf Deine Worte,
denn sie werden Handlungen.*

*Achte auf Deine Handlungen,
denn sie werden Gewohnheiten.*

*Achte auf Deine Gewohnheiten,
denn sie werden Dein Charakter.*

*Achte auf Deinen Charakter,
denn er wird Dein Schicksal.*

aus dem Talmud

Danksagung

Mein herzlicher Dank gebührt Dr. Heiko Lickert für die Möglichkeit, meine Promotion als erste Doktorandin in seiner jungen Arbeitsgruppe durchführen zu können. Die immerwährende positive Einstellung zu seiner Arbeit, der schier unerschöpflichen Enthusiasmus und die ausgeprägte Begeisterungsfähigkeit können einem nur ein Vorbild sein. Vielen Dank für das entgegengebrachte Vertrauen, die Hilfe und Unterstützung, anregende und unterhaltsame Diskussionen!

Ich möchte Professor Wolfgang Wurst für die Übernahme der Betreuung der Arbeit und PD Dr. habil. Johannes Beckers für die Teilnahme an der Prüfungskommission für diese Arbeit danken. Zudem gilt mein Dank Professor Gierl für unkomplizierte und selbstverständliche Übernahme des Vorsitzes der Prüfungskommission.

Mein Dank geht an die gesamte Arbeitsgruppe für die gute Zusammenarbeit über Jahre hinweg und die vielen gemeinsamen Unternehmungen; dies gilt insbesondere für meine allerersten Kolleg:en: vielen Dank an Dr. Ingo Burtscher, Wenke Barkey und Daniela Wagner für die Unterstützung und die gute Zusammenarbeit, für viele unterhaltsame Stunden und manches Gespräch und Freundschaft, die mir als Norddeutsche das Einleben im Süden doch sehr erleichtert haben.

Darüber hinaus möchte ich mich bei allen meinen Mitdoktoranden und -doktorandinnen bedanken, die mich mehr oder weniger lange auf die ein oder andere Weise begleitet haben - ein bißchen Ironie kann einem das Leben manchmal deutlich erleichtern! Danke für unterstützende Worte, das entgegengebrachte Verständnis und die wertvollen zwischenmenschlichen Erfahrungen an (nach zeitlichem Erscheinen - „in order of appearance") Doris Bengel, Sascha Imhof, Moritz Gegg, Marion Vollmer und Perry Liao. Perry sei an dieser Stelle auch für etliche gemeinsame Tassen Kaffee und die Korrekturlesearbeiten an dieser Arbeit gedankt! Für ein bißchen mehr Internationalität in der Arbeitsgruppe, belgische Schokolade und kritische, hilfreiche Kommentare zu verfaßten Texten geht mein Dank an Dr. Claude van Campenhout, für das gewisse italienische Etwas an Patrizia Giallonardo.

Vielen Dank auch an Dr. Ursula Gaio für viele sehr nette Abende und vor allem den Einsatz als wandelndes PubMed...

Meinen zwei äußerst gut laufenden „Kooperationen", Dr. Susanne Neschen und Dominik Lutter danke ich herzlich für die gute Zusammenarbeit und die vielen interessanten Einblicke in andere Arbeitsbereiche - würden doch nur alle Kooperation so gut laufen!

Gute Resultate im Hinblick auf die Mausversuche habe ich auch der Unterstützung der Tierpfleger zu verdanken - vielen Dank für das Umsorgen und Mitbeobachten meiner Mäuse!

Viele Freunde und Bekannte haben mein Leben in den letzten dreieinviertel Jahren gekreuzt und mich durch einen Lebensabschnitt begleitet, der durch unglaublich viele Veränderungen und wichtigen Erfahrungen gekennzeichnet war - sie seien an dieser Stelle nicht unerwähnt, wenn auch nicht alle namentlich aufgeführt.

Ich möchte allerdings an dieser Stelle (stellvertretend) Jennifer Heinke und Philipp Eßer erwähnen, deren Freundschaft schon wie taus länger währt und der auch die größer gewordene Entfernung nichts anhaben kann! Es ist schön, wenn man einen Menschen kennt, der einen ohne viele Worte verstehen kann.

Lieber Dominik, auch Dir möchte ich an dieser Stelle danken, vor allem für das Verständnis und die unglaubliche Ruhe, die Du mir und meiner Arbeit entgegengebracht hast.

„Last but not least" soll hier der größte Dank an meine Eltern gehen: vielen Dank für die Unterstützung in all den Jahren, ob nun für geschickte Schokoladen-Notrationen für die „Endphase" oder wichtige grundlegende statistische Informationen in einer Geschwindigkeit, die Google in den Schatten stellt. Schön, daß es Menschen gibt, die immer an mich glauben! Danke!

Contents

1. Abstract		13
2. Introduction		15
2.1.	Early embryonic development in the mouse	15
2.2.	Formation of the endoderm and the development of the endodermal organs	18
2.3.	Important transcription factors during endoderm development: *Foxa2*	21
2.4.	Important transcription factors during endoderm development: *Sox17*	22
2.5.	Stem cells – a model for embryonic development and an indispensible tool to analyze gene function	23
2.6.	Alterations of the genome	25
2.7.1.	MiRNAs in development and differentiation	26
2.7.2.	The miRNA pathway: from transcription to target interaction	28
3. Rationale		31
4. Results		33
4.1.	Cell lineage tracing - Generation and analysis of Cre mouse lines	34
4.1.1.	Targeting of the *Sox17* locus	34
	a) Design and generation of the targeting vector for the *Sox17* locus	34
	b) Targeting of ES cells for homologous recombination and deletion of the selection cassette for the *Sox17* locus	35
4.1.2.	Analysis of the $Sox17^{iCre/+}$ mouse line	35
4.1.3.	Recombination activity of the Sox17-iCre recombinase at early embryonic stages and after birth	37
4.1.4.	Targeting of the *Foxa2* locus	37
	a) Design and generation of the targeting vector for the *Foxa2* locus	37
	b) Targeting of ES cells for homologous recombination for the *Foxa2* locus	38
4.1.5.	Analysis of the $Foxa2^{iCre/+}$ mouse line	38
4.1.6.	Recombination activity of the Foxa2-iCre recombinase at early embryonic stages	38
4.1.7.	Recombination activity of the Foxa2-iCre recombinase in embryonic organs	39
4.2.	Characterization of the $Foxa2^{iCre}$ hypomorphic allele	41
4.3.	Analysis of the metabolism of $Foxa2^{iCre\Delta neo/iCre\Delta neo}$ mice	44
4.4.	Conditional knock-out analysis	46
4.4.1.	Characterization of the β-catenin knock-out in the *Foxa2*-positive cell population: $Foxa2^{iCre/+}$; $β$-$catenin^{flox/flox}$; $R26^{R/+}$ mice	46
4.4.2.	Analysis of the metabolism of $Foxa2^{iCre\Delta neo/+}$; $β$-$catenin^{flox/flox}$; $R26^{R/+}$ mice	47
4.4.3.	Characterization of the ß-catenin knock-out in the *Foxa2*-positive cell population: $Foxa2^{iCre/+}$; $β$-$catenin^{floxdel/flox}$; $R26^{R/+}$ mice	48
4.5.	Differentiation of ES cells into endoderm	51
4.5.1.	Establishment of an *in vitro* differentiation system from ES cells into endoderm	52
4.5.2.	Characterization of the *in vitro* differentiation system	54

Contents

4.5.3.	*In vitro* differentiation – onset of marker expression (FACS analysis)	54
4.5.4.	*In vitro* differentiation – onset of marker expression (live imaging)	55
4.6.	Using the *in vitro* differentiation system to screen for micro RNAs influencing endoderm development	56
4.6.1.	Generation of the miRNA expression vector	57
4.6.2.	Design of miRNA expression constructs	58
4.6.3.	Fluorescent analysis of the miRNA-transgenic ES cell clones using *in vitro* differentiation and immunostaining	60

5. Discussion   67

5.1.	The generation of iCre mouse lines under the control of two endodermally expressed transcription factors	67
5.2.	$Sox17^{iCre/+}$ mice – a valuable tool for specific deletion of genes in arteries and an indication of a second promoter activated in the endoderm	68
5.3.	The $Foxa2^{iCre/+}$ mouse line – a new tool to analyze gene function in a variety of tissues	70
5.4.	The $Foxa2^{iCre}$ allele, a hypomorph with interesting potential	72
5.5.	Conditional deletion of β-catenin in the domain of Foxa2-iCre recombination activity	74
5.6.	Establishment and characterization of an *in vitro* ES cell assay – steps towards an in vitro system for embryonic development	76
5.7.	Fusion proteins as miRNA-sensors – a tool to test micro RNAs *in vitro*?	77

6. Material and Methods   83

6.1.	Material	83
6.1.1.	Equipment	83
6.1.2.	Consumables and kits	84
6.1.3.	Chemicals	85
6.1.4.	Buffers and solutions	87
6.1.5.	Enzymes and enzyme kits	89
6.1.6.	Antibodies and sera	90
6.1.7.	Vectors and BACs	90
6.1.8.	Oligonucleotides	90
6.1.9.	Probes	92
6.1.10.	Molecular weight markers	92
6.1.11.	Bacteria and culture media	93
6.1.12.	Cell lines, culture media and solutions	93
6.1.13.	Mouse lines	94
6.2.	Methods	94
6.2.1.	Methods in molecular biology	94
6.2.1.1.	Preparations of nucleic acids: DNA preparations	94
	a) Plasmid and BAC preparations	94
	I. Plasmid preparations according to the QIAGEN Plasmid Kits	94
	II. BAC mini preparation according to Copeland	94
	III. BAC maxi preparation according to the NucleoBond BAC Purification Maxi Kit	95
	b) Preparation of genomic DNA from cells or tissue	95
	I. Isolation of genomic DNA from cells in 96-well plate format	95

	II. Isolation of genomic DNA from cells in cell culture dish format	95
	II. Isolation of genomic DNA from mouse tail biopsy	96
6.2.1.2.	Preparations of nucleic acids: RNA preparations	96
	a) Preparation of total RNA according to the QIAGEN RNA Mini Kit	96
6.2.2.	Determination of the concentration of DNA and RNA solutions	96
6.2.3.	Reverse trancription	97
	a) DNAse digest of RNA samples	97
	b) RNA precipitation with LiCl	97
	c) Reverse transcription with oligo dT-primers	97
	d) PCR on cDNA	97
6.2.4.	Restriction analysis of DNA	98
	a) Analytical or preparative restriction digest of plasmid or BAC DNA	98
	b) Restriction digests of genomic DNA	98
	I. Restriction digests of genomic DNA from (ES-) cells and tissue for Southern blot analysis	98
	II. Restriction digests of genomic DNA from (ES-) cells in 96-well plates for Southern blot analysis	99
6.2.5.	Gelelectrophoresis	99
	a) Analytical agarose gelelectrophoresis	99
	I. DNA	98
	II. RNA	99
	b) Preparative agarose gelelectrophoresis	99
6.2.6.	Generation of blunt ends	99
6.2.7.	Dephosphorylation of linearised DNA	100
6.2.8.	Ligation	100
6.2.9.	Cloning of short DNA sequences using complementary DNA oligonucleotides	100
	a) Hybridisation	101
	b) Phosphorylation	101
	c) Ligation of hybridised oligonucleotides	101
6.2.10.	Generation of competent bacteria	101
	a) Generation of electro-competent bacteria (*E. coli* K-12 XL1-Blue)	101
6.2.11.	Transformation of bacteria	102
	a) Transformation of bacteria using electroporation	102
	b) Transformation of bacteria using heat shock	102
6.2.12.	Bacterial homologous recombination	102
6.2.13.	DNA sequencing	103
	a) Sequencing reaction	103
	b) DNA preparation for sequencing: ethanol – sodium acetate precipitation	103
6.2.14.	Southern Blot	104
	a) Gel electrophoresis	104
	b) Blot	104
	c) Hybridisation	104
	I. Prehybridisation	104
	II. Preparation of samples – radioactive labelling of the probe	104
	III. Hybridisation	105
	d) Preparation of samples - radioactive labelling of the probe according to Roche	105
6.2.15.	Detailed description of the generation of the targeting construct for *Sox17*	105
6.2.16.	Detailed description of the generation of the targeting construct for *Foxa2*	106
6.2.17.	Detailed description of the generation of the vector for miRNA expression	106
6.2.18.	Generation of miRNA overexpression vectors	106

6.3.	Methods in protein biochemistry	107
6.3.1	Extraction of proteins	107
	a) Protein extraction: whole cell lysat	107
	b) Protein extraction: nucleic proteins according to the CelLytic™ NuClear™ Extraction Kit (Sigma NXTRACT)	107
	I. Extraction from tissue (Embryo till E9.5)	107
	II. Extraction from cells with detergent (ES cells, differentiated)	107
6.3.2.	Determination of protein concentrations	107
	a) Determination of protein concentrations using BCA	107
	b) Determination of protein concentrations using Bradford	108
6.3.3.	Western Blot	108
	a) Denaturing SDS-polyacrylamid gelectrophorese	108
	b) Immunoblot: Semi-dry Blot	109
	c) Immunostaining	109
6.3.4.	Immunostainings	109
6.4.	Methods in cell biology	109
6.4.1.	(ES) Cell culture	109
	a) Culture of MEF	110
	b) Treatment of MEF with mitomycin C (MMC)	110
	c) Seeding of MEF for ES coculture	110
	d) Thawing of ES cells	110
	e) Passaging of ES cells	110
	f) Cryoconservation of ES cells	110
6.4.2.	Homologous recombination in ES cells	111
	a) Transformation of ES cells by electroporation	111
	b) Picking of ES cell clones	111
	c) Expansion of ES cell clones	111
	d) Cryoconservation of ES cell clones in 96-well-plates	111
	I. Generation of $Sox17^{iCre/+}$ cells	112
	II. Generation of $Foxa2^{iCre/+}$ cells	112
	III. Generation of miRNA-transgenic ES cell clones	112
6.4.3.	*In vitro* differentiation of ES cells in *Foxa2*- and *Sox17*-positive progenitor cells	112
6.4.4.	Ca/phosphate transfections	112
6.4.5.	Fluorescence activated cell sorting analysis: FACS	112
6.5.	Methods in embryology	112
6.5.1.	Mouse husbandry and -matings	112
6.5.2.	Genotyping of mice and embryos using PCR	113
	a) Genotyping of $Sox17^{iCre/+}$ mice	113
	b) Genotyping of $Foxa2^{iCre/+}$ mice	113
	c) Genotyping of R26R mice	113
	d) Genotyping of $β\text{-}catenin^{flox/+}$ or $β\text{-}catenin^{floxdel/+}$ mice	113
	e) Genotyping of *Flp-e* mice	114
6.5.3.	Isolation of embryos and organs	114
6.5.4.	X-gal (5-bromo-4-chloro-3-indolyl β-D-galactoside) staining	114
6.5.5.	Clearing of embryos and organs	114
6.6.	Methods in histology	114
6.6.1.	Paraffin sections	114
	a) Embedding of embryos and organs for paraffin sections	114

	b) Sectioning of paraffin blocks	114
	c) Histological staining of paraffin sections using Nuclear Fast Red (NFR)	114

7. References 117

8. Appendix 133

8.1.	Publications	133
8.2.	List of abbreviations	133
8.3.	Listing of figures	135
8.4.	Listing of tables	137
8.5.	Listing of charts	137

1. Abstract

During gastrulation in the mouse the three principle germ layers form: ectoderm, mesoderm and endoderm. The endoderm gives rise to the endoderm-derived organs, such as thyroid, thymus, lung, liver, stomach and gastrointestinal tract. However, little is known about the signals and molecules that regulate the first lineage decisions in the endoderm. Moreover, the progenitor populations are yet to be identified and it is important to identify subpopulations that give rise to specific endodermal organs. The forkhead transcription factor *Foxa2* and the SRY-related HMG box transcription factor *Sox17* are among the first marker genes in the endoderm. They are indispensible for endoderm formation and development. However, it is still not known to which organs *Foxa2* and/or *Sox17* positive progenitor cells contribute to and how this lineage segregation is achieved on a molecular level.

As a first step to answer these questions, two Cre mouse lines were generated that express Cre under the control of these two transcription factors. They were used to genetically lineage trace these endoderm subpopulations and to clarify the involvement of pathways that are essential for the differentiation and specification of cells within the endoderm using conditional knock-out analysis.

Unexpectedly, the lineage tracing of *Sox17* revealed that the expression of *Sox17*-iCre is restricted to the endothelium of arteries and shows little expression in endoderm-derived organs. These results revealed that *Sox17* transcription is regulated from independent promoters which drive expression in the vascular endothelium, in the blood stem cell system and in the endoderm germ layer. The arterial endothelia and hematopoietic stem cell specific inactivation of *Sox17* might be an useful tool to study gene function in these lineages. Additionally, the *Sox17*-iCre mouse line with specific Cre recombination activity in arterial endothelium will be a useful tool for studies of embryonic and adult arterial development. It clearly shows differential activation of *Sox17* in artery vs. vein endothelium.

Genetic lineage analysis for the *Foxa2*-iCre mouse line revealed that *Foxa2* positive cells are the progenitor population for all endoderm-derived organs and suggests that *Foxa2* is a marker for a very early progenitor population, possibly an endodermal stem cell population. Moreover, using an exon 1 knock-in strategy revealed a second downstream promoter active in the endoderm lineage. Due to this promoter we generated a hypomorphic allele of *Foxa2* which is very useful to analyze *Foxa2* gene function in metabolism and is most likely useful for the analysis of Parkinson's disease.

Besides the analysis of endoderm differentiation *in vivo* this exposition also deals with *in vitro* endoderm differentiation. *In vitro* differentiations of stem cells in regard to cell replacement therapies have made enormous progress during the last decade. Since terminally differentiated cells can be dedifferentiated into a pluripotent state (iPS cells = induced pluripotent stem cells) one might also be able to circumvent the ethical problem of human embryonic stem cells and *in vitro* differentiation systems will gain more importance.

Here, the establishment and characterization of an efficient endoderm *in vitro* differentiation protocol with many possible applications is described. In this thesis the robust differentiation technique allowed for the testing of the impact of different miRNAs on endoderm formation *in vitro*. Cooperation of predictive information gained on the basis of bioinformatic screens and experimental approaches (based on the differentiation of stable transgenic miRNA ES cells clones) could identify two miRNAs, miR335 and miR194, that negatively and positively regulate *Foxa2*-expression, respectively.

2. Introduction

2.1. Early embryonic development in the mouse

Embryogenesis begins when penetration by the sperm leads to the fertilization of the oocyte at what is then termed embryonic day 0 (E0). From this point onwards, the fertilized oocyte is called a zygote (see figure 1). Immediately after fertilization the zygote that is surrounded by the protecting *zona pellucida* starts dividing and forms the morula. The morula is a spheric cluster of 4-16 cells, called blastomeres, and has the same total volume as the original zygote. Subsequently, the cells inside and outside of the morula are determined to undergo opposing differentiation processes. Cells of the outer morula form the trophoblast or trophectodermal cells, the outer layer of the blastula stage embryo, while the cells from the inside of the morula form the inner cell mass (ICM) and the overlying primitive endoderm within the blastocyst. The primitive endoderm forms the border of the ICM to the blastocoel cavity (for review see Wang and Dey, 2006). Embryonic stem (ES) cells can be isolated from the ICM at this stage (Martin, 1981).

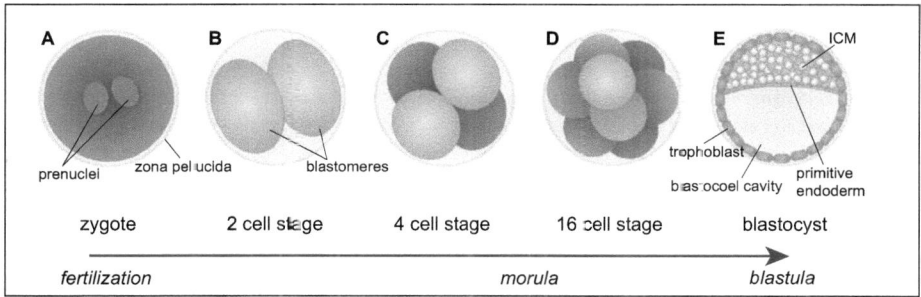

Figure 1: The first steps in embryonic development – reaching the blastula stage

Right after fertilization the zygote is surrounded by the protecting *zona pellucida* (A). Shortly after fertilization the zygote starts dividing (B) and forms the morula (C; D). When the embryo has reached the blastula stage (E) it consists of an outer layer, the trophoblast (grey), the ICM (lilac) and the primitive endoderm (orange) that forms the border between the ICM and the blastocoelic cavity (yellow). (Abbreviations: ICM = inner cell mass)

With the formation of the blastocyst the first embryonic axis is already defined (embryonic-abembryonic or proximal-distal axis; see figure 2). At about E3.5-4.5 the blastocyst first hatches out the *zona pellucida* and then implants into the receptive maternal uterus. Afterwards cells of the trophoblast interact with cells of the uterus mucosa, penetrate the luminar epithelium and form a tissue that combines maternal and (extra-) embryonic tissue: the placenta (for review see Wang and Dey, 2006).

As the embryo develops further, the ICM elongates into the blastocoel cavity, forming a cup-shaped embryo consisting of an extra-embryonic and an embryonic part, the latter will eventually give rise to the embryo proper. The embryonic component is comprised of the embryonic epiblast, a one cell layer thick columnar epithelium, that is surrounded by the visceral endoderm (VE), a planar epithelium that is part of the extra-embryonic tissues, contributing only to the yolk sac (for review see Beddington and Robertson, 1998 and 1999; Ang and Constam, 2004).

The first morphological changes overall can be detected in the extra-embryonic portion of the embryo, when the distal VE (DVE) forms at E5.5 (Beddington and Robertson, 1999) and moves anteriorly. The DVE is a thickening in the VE at the distal tip of the embryo that, once moved to the anterior side, is termed anterior VE (AVE formation at E6.0, see figure 3; Srinivas et al., 2004). The columnar DVE, and later on the AVE, is also molecularly marked by the expression of the transcription factor *Hex* (haematopoetically expressed homeobox gene; Thomas et al., 1997; Thomas et al., 1998). It was shown, however, that other markers, like *Otx2*, *Lim1*, *Gsc*, *Cerl* and *Nodal*, *Cripto*, *Wnt3* and *Bmp4*, in the VE and the extra-embryonic and embryonic epiblast, respectively, are expressed

Introduction

Figure 2:
Embryonic development of the mouse until implantation
At E4.5 the blastocyst has hatched out the zona pellucida. The polar trophectoderm (blue) interacts with the maternal uterus and implants (E5.5). The distal VE (red) can be detected as a thickening at the distal tip of the epiblast (lilac) that has elongated into the blastocoelic cavity, formed by the cell of the ICM (E4.5; lilac). At E5.75 the distal VE starts to move to the future anterior side, now called the AVE (red). (Abbreviations: AVE = anterior visceral endoderm; E = embryonic day; ICM = inner cell mass; VE = visceral endoderm)

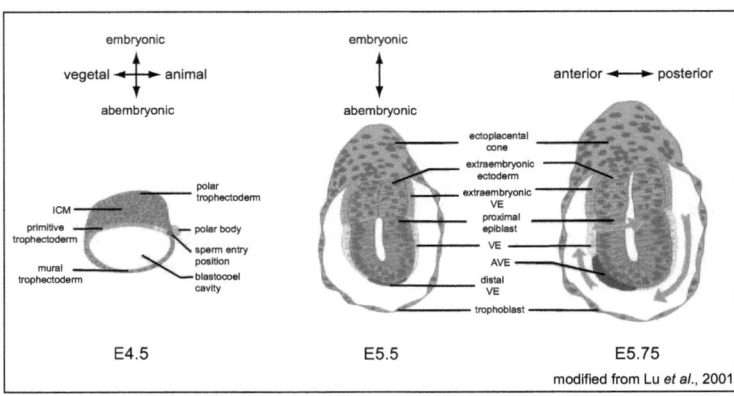

modified from Lu et al., 2001

asymmetrically earlier than E6.5 (Ang and Rossant, 1994; Belo et al., 1997; Kimura-Yoshida et al., 2007; Shawlot et al., 1998) thereby defining to anterior-posterior axis.

The next crucial step in early mouse embryonic development is the process of gastrulation (see figure 3). Generally, during gastrulation in vertebrates and consequently in mice, the three principle germ layers: ectoderm, mesoderm and endoderm, are formed by cells coming from the epiblast (from the inside outwards) (for review see Beddington and Robertson, 1998 and 1999). Gastrulation in the mouse begins at E6.5 when the cells which will give rise to endoderm and mesoderm leave the epiblast and undergo an epithelial-mesenchymal transition at the posterior side of the embryo. Within those cells giving rise to the embryo proper the formation of the primitive streak (starting at E6.5) is the first visible indication of the second embryonic axis (anterior-posterior axis). These differentiation processes do not only result in a change of morphology of the cells involved but also give rise to a transient structure, called the primitive streak, which originates at the future posterior side of the embryo along the embryonic-extra-embryonic border. It has been shown that one function of the AVE is to repress posterior gene expression in the anterior epiblast (Kimura et al., 2000; Perea-Gomez et al., 2001) causing gastrulation to appear and the primitive streak to form at the opposite side of the embryo (Perea-Gomez et al., 2002). Dkk1 (Dickkopf; Glinka et al., 1998), Lefty1/2 (left right determination factor; Perea-Gomez et al., 2002) and Cerl 1 (Cerberus-like 1; Piccolo et al., 1999) are expressed in the AVE and repress TGFβ and Wnt signals from the posterior side and thereby define the anterior side. The posterior epiblast is marked by the expression of Nodal and Wnt3. Mutants with deletion of one of both genes do not form a primitive streak (Conlon et al., 1994; Liu et al., 1999).

It has been postulated that a possible intermediate mesendodermal precursor cell population, a cell type that forms both mesoderm and endoderm, exists in the mouse due to the fact that a bipotent precursor cell has already been identified in zebrafish, Caenorhabditis elegans and Xenopus (Rodaway et al., 1999; Kimelman and Griffin, 2000; Rodaway and Patient, 2001). Additionally, it has been shown that at least in vitro there is a cell population that can give rise to both mesoderm and endoderm (Tada et al., 2005). The next process is the reverse transition from mesenchyme to epithelium when the newly formed definitive endoderm (DE) replaces the overlying VE, while the mesodermal cells stay in their mesenchymal state between the ectoderm (cells that remain in the epiblast) and the definitive endoderm.

At the distal end of the primitive streak the node forms (see figure 3). The node is one of the cell populations having organizer function in the late gastrula stage embryo (Blum et al., 1992; Beddington, 1994). The organizer, by definition, is a cell population capable of inducing an embryonic axis as shown by Spemann and Mangold from transplantation experiments of the blastopore lip in the newt (Spemann organizer; Spemann and Mangold, 1924). Later, analogous embryonic signalling centres with organizer function and similar fates of participating cells could be identified in other vertebrates like the fish (Shih and Fraser, 1996), the chick (Hensen's node; Hensen, 1876; Wetzel, 1925), the rabbit (Waddington, 1933) and the mouse (Kinder et al., 2001; Beddington, 1994; Tam et al.,

1997). In the mouse Kinder et al. (2001) could show that the organizer consists of a dynamic population of cells that have a different fate and varying potential to induce secondary axes, depending on the stage of gastrulation (Kinder et al., 2001). They showed that cells with organizer function in the early streak embryo (early streak or gastrula organizer; EGO) form a fraction of the most anterior definitive endoderm (ADE) and prechordal plate (PCP) mesoderm, while the organizer of the mid-gastrula stage embryo (mid gastrula organizer; MGO) gives rise to part of the notochord, ADE and axial mesoderm. The late gastrula organizer (LGO), also called the node, however, participates in the formation of the notochord and floor plate in more posterior parts of the embryo (Kinder et al., 2001).

While gastrulation proceeds, the primitive streak elongates to the distal tip of the embryo. At the end of this differentiation process (at E7.5) the mesoderm and the endoderm surround the whole embryo and the DE has completely replaced the VE. At this point, the DE is a one cell layer thick sheet of approximately 500 flattened cells (for review see Wells and Melton, 1999).

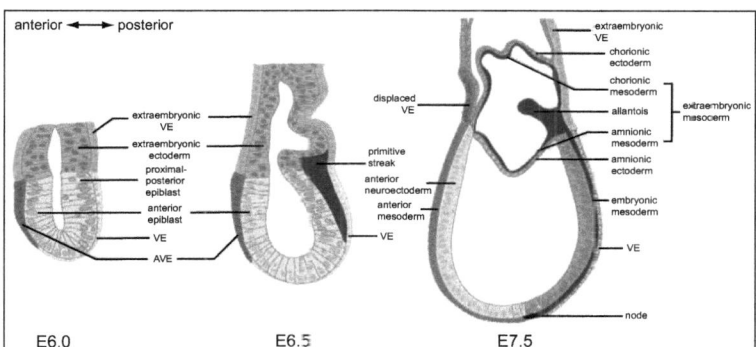

Figure 3:
Gastrulation in the mouse
At E6.0, shortly before gastrulation occurs, the posterior-anterior axis is already defined by the AVE (red). At E6.5, the primitive streak (violet) starts to form at the opposite side, the posterior side of the embryo. Cells of the epiblast (light pink) undergo an epithelia-mesenchymal transition and migrate out the primitive streak. They become either DE (green/yellow) and replace the overlying VE or they become mesoderm and stay between the epiblast, now called ectoderm, and the DE. Gastrulation proceeds until E7.5 when the node (the late gastrula organizer) has reached the tip of the gastrulating embryo. Abbreviations: AVE = anterior visceral endoderm; DE = definitive endoderm; E = embryonic day; VE = visceral endoderm) modified from Lu et al., 2001

After gastrulation the headfolds materialize at the anterior side of the embryo (early headfold ~E7.75; see figure 4). With the formation of the headfolds and the beginning of somitogenesis (the generation of the somites from the presomitic mesoderm, further explained below) the anterior intestinal port (AIP) forms, a fold in the anterior most DE that moves posteriorly. A similar fold arises later in the most posterior part of the DE from cells moving anteriorly which form the caudal intestinal port (CIP). The two folds meet at the stalk of the yolk sac. This way, the most anterior and posterior parts of the DE form the ventral gut tube and all organs that will arise from that, while the midline endoderm gives rise to the dorsal part of the gut tube and the associated organs, respectively (for review see Lewis and Tam, 2006). Around E8.5 to E9.0, while the gut closes completely, the embryo turns so that the dorsal side faces the outside and the ventral side lies inside (see figure 4).

Figure 4: Turning of the mouse embryo
After gastrulation the endoderm is facing the outside of a mouse embryo (E7.75). Migration events during somitogenesis from two pockets of endoderm: the anterior intestinal port and the caudal intestinal port (E8.5). As the gut closes completely the embryo is forced to turn between E8.5 and E9.5. At E9.5 the gut has closed and the endoderm is fully ventralized.

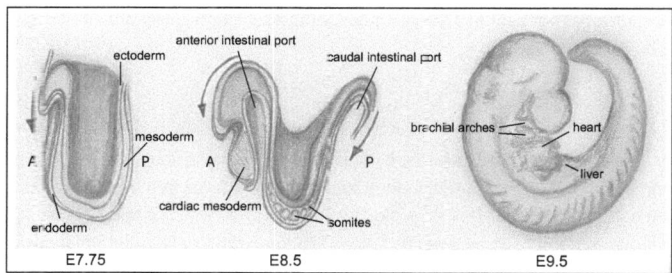

Introduction

Originating from the embryonic gut tube the endoderm will form all its derived organs during organogenesis from E9.0 onwards (for review see Wells and Melton, 1999), namely the thymus, the thyroid, the epithelium of the trachea, the pharynx, the lung, the liver, the pancreas, the stomach, the gastrointestinal tract, the epithelial part of the ureter and the bladder (see figure 5).

After gastrulation the mesoderm can be divided into four major groups: the axial, the paraxial, the intermediate and the lateral (plate) mesoderm. The axial mesoderm along the midline will give rise to the *chorda dorsalis* while the paraxial or presomitic mesoderm will segregate into the somites, beginning anteriorly immediately caudal to the otic vesicle and proceeding posteriorly on both sides of the neural tube and the notochord to the caudal tip of the embryo. The somites can be subdivided into the ventral sclerotome which will later differentiate into cartilage and bones and the dorsal dermomyotome (including dermatome and myotome; Hirsinger *et al.*, 2000), which gives rise to skeletal muscles, supporting and connective tissues, and the dermis of the back. The connection of the somites and the lateral plate mesoderm arises from the intermediate mesoderm that will also form the urogenital apparatus, including the gonads and defined nephronal structures (defined cells of the kidneys). The lateral mesoderm is subdivided into the splanchnic and the somatic layer. While the splanchnopleuric mesoderm lies ventrally and is associated with the endoderm, the somatopleuric mesoderm forms a dorsal layer associated with the ectoderm. Both layers take part in the formation of the peritoneum; however, the splanchnic mesoderm gives rise to the future wall of the gut and the whole circulatory system including the heart, while the somatic layer only takes part in the establishment of the lateral and ventral walls of the embryo. The lymphatic system is also derived from the mesoderm and forms by budding from the veins. It is derived from a mixture of paraxial and lateral mesoderm (for review see Hong *et al.*, 2004). Hematopoietic cells are formed by cells coming from the posterior mesoderm. The hematopoietic stem cells are either formed in the yolk sac (starting at E7.5), in the para-aortic splanchnopleura (E8.5-9.5) or in the aorta-gonad-mesonephros (AGM) region (E10.5-11.5) till the production finally shifts to the embryonic liver and later, around birth, to the bone marrow (for review see Baron, 2003). (For derivatives of the mesoderm see figure 5)

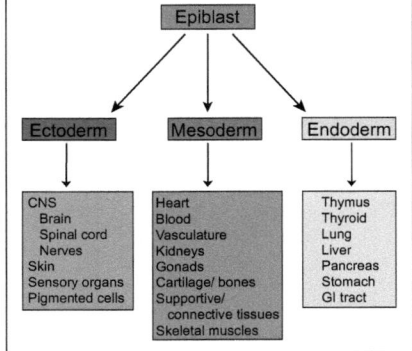

Figure 5:
The three principle germ layers and their derivatives
Pluripotent cells of the epiblast are capable of forming cells of all three germ layers: ectoderm, mesoderm and endoderm (dark boxes). Over the course of development each of the three germ layers will give rise to certain tissues in the body (light boxes). (Abbreviations: CNS = central nervous system; GI = gastrointestinal)

The ectoderm folds up to build the neural folds first, allowing a cavity in the neural plate. The folds will then move dorsally towards each other and their fusion leads to the closure of the cavity and release of the neural tube from the rest of the ectoderm. In the course of development the ectoderm gives rise to all neuronal structures (brain, spinal cord and nerves in the periphery of the body), sense organs and the skin, as well as the inversions of the epithelium at both ends of the gut tube that build the oral cavity and the anus, and every kind of pigmented cells (see figure 5).

2.2. Formation of the endoderm and the development of the endodermal organs

At the end of gastrulation and the beginning of the headfold stage at E7.5 the endoderm – in contrast to mesoderm and ectoderm – is not determined in regard to its regional fate demonstrated by the fact that the anterior half of the endoderm can still be respecified when associated with posterior tissues (Wells and Melton, 2000). However, endoderm does become specified due to its localization within the anterior-posterior axis (Lawson *et al.*, 1991; Wells and Melton, 2000; Horb and Slack, 2001; see figure 6), proven by asymmetric expression of several factors, including anterior markers like *Cerl* (*Cerberus like 1*; Bouwmeester *et al.*, 1996), *Hesx1* (*homeo box gene expressed in ES cells*; Thomas and Beddington, 1996), *Otx1* (*orthodenticle homolog 1*; Rhinn et al., 1998) in the foregut region and posterior markers *IFAB-P* (*intestinal fatty acid binding protein*; Green et al., 1992) and *Cdx2* (*caudal type homeo box 2*; Beck et al., 1995) in the hindgut region. To establish foregut versus hindgut endoderm

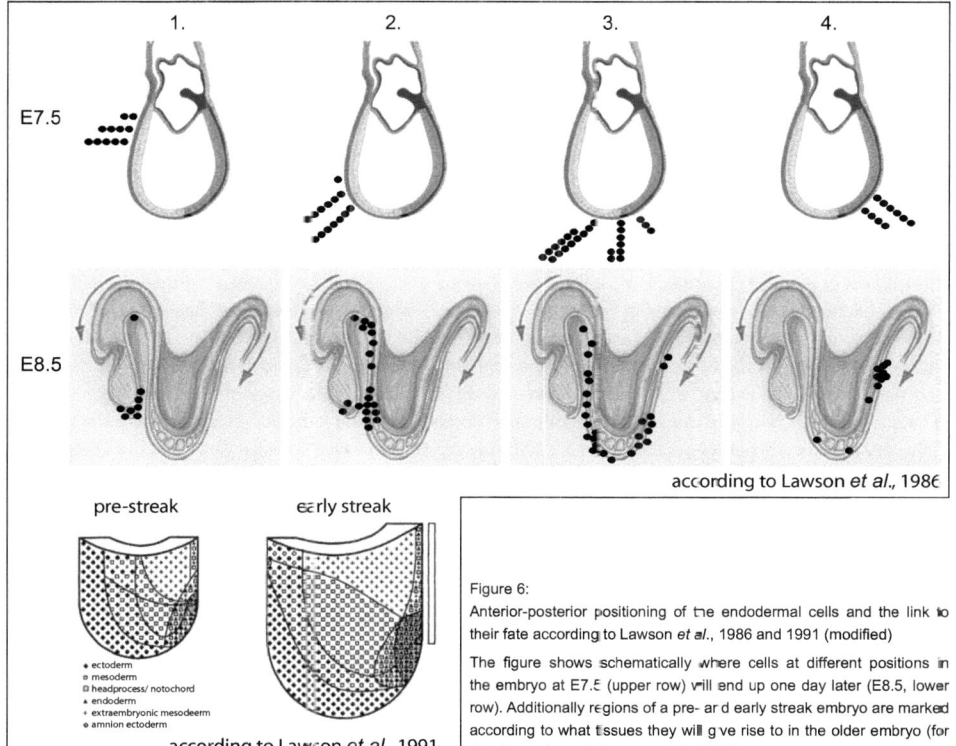

Figure 6:
Anterior-posterior positioning of the endodermal cells and the link to their fate according to Lawson et al., 1986 and 1991 (modified)

The figure shows schematically where cells at different positions in the embryo at E7.5 (upper row) will end up one day later (E8.5, lower row). Additionally regions of a pre- and early streak embryo are marked according to what tissues they will give rise to in the older embryo (for details see legend; Lawson et al., 1991).

a restriction of fibroblast growth factor (Fgf) and Wnt signalling to the posterior side in the gastrula stage embryo is necessary and the exclusion of these posteriorizing factors is required for the proper development of the foregut (Dessimoz et al., 2006; McLin et al., 2007).
Adjacent germ layers in close proximity to the endoderm, can also express a variety of signalling molecules, such as Fgf from the mesoderm that have been shown to induce liver and lung development in a dose-dependent manner *in vitro* and *in vivo* (Serls et al., 2005; Dessimoz et al., 2006). The Wnt and retinoic acid (RA) signalling pathways, in addition to the hedgehog pathway have also been implicated in foregut formation and patterning (McLin et al., 2007; for review see Wodarz and Nusse, 1998; Ross et al., 2000; Nederreither and Dollé, 2008). These factors are known to regulate expression of important transcription factors, like Foxa (forkhead box a), ParaHox (para homeo box) and Hox (homeo box) genes, mediators of cell fate (Grapin-Botton, 2005). It therefore suggests that spatially and temporally limited signals from adjacent tissues pattern and determine the endoderm after gastrulation in a concentration-dependent manner and that these signals are instructive rather than permissive (Well and Melton, 1999).

During organogenesis (starting around E9.0) the anlage of endodermal organs are first formed by buds from the primitive gut tube into the mesodermal periphery (for review see Wells and Melton, 1999). The endodermal cells therefore migrate out of the tube and invade the mesoderm (see figure 7).
Although the later development of the endoderm-derived organs has been well studied, the signals and signal

Introduction

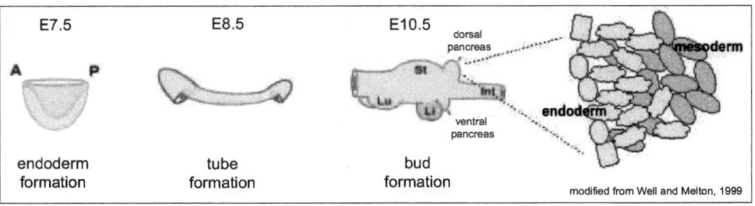

Figure 7: The embryonic gut – budding of the organs

The figure shows the endoderm between E7.5 and E10.5 (yellow). At E7.5 the endoderm covers the outside of the mouse embryo. Around E8.5 the gut tube forms, starting with the two gut pockets at the anterior and posterior ends. At E10.5 the endoderm derived organs bud from the gut tube. The endodermal cells (yellow) therefore invade the neighbouring mesoderm (blue) forming the organ anlagen.
(Abbreviations: Int = intestine; Li = liver; Lu = lung; St = stomach)

transduction pathways that play important roles in the early lineage decisions and steps of differentiation have not been well-characterized and still remain unclear. It is already known that the *TGFβ* family member *Nodal* is necessary to establish the anterior-posterior axis by restricting gastrulation to the posterior side of the embryo (Conlon et al., 1994; Varlet et al., 1997; Brennan et al., 2001; Perea-Gomez et al., 2002). A Nodal/TGFβ gradient is sufficient to induce endoderm versus mesoderm in the gastrulating embryo, with high and low levels respectively (Zhou et al., 1993; Conlon et al., 1994; Alexander and Stainier, 1999; Tremblay et al., 2000; Lowe et al., 2001; Vincent et al., 2003). There is also evidence for a gradient of canonical Wnt-signalling (see figure 8) being relevant to endoderm and mesoderm determination in the same manner. Embryos lacking a key factor of canonical Wnt-signalling, β-catenin, in the *K19-Cre* (*Cytokeratin 19*) expression domain show the formation of ectopic mesodermal cells within the endodermal germ layer (Lickert et al., 2002). β-catenin gets activated when Wnt binds to its cell surface receptor of the Frizzled (FRZ) family. Binding to FRZ Wnt leads to the release of Dishevelled (Dsh) from the Wnt receptor complex. This activated form of Dsh can block the proteolytic degradation complex of β-catenin including axin, GSK-3 (glycogen synthase kinase 3) and APC (adenomatosis polyposis coli). If the degradation complex is blocked β-catenin accumulates and shuttles to the nucleus where it can promote transcription of specific genes in concert with TCF/LEF (T-cell factor/lymphoid enhancer factor) transcription factors (see figure 8).

To determine which signalling pathways play roles in which differentiation processes it is necessary to be able to

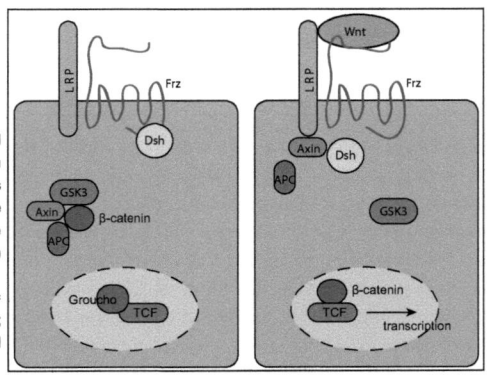

Figure 8: Canonical Wnt signalling

In the absence of Wnt (left picture) Dsh (yellow) is bound to Frz (red line) and can therefore not block the degradation complex β-catenin (red circle) is bound to. When Wnt (light blue-green) binds to its surface receptor Frz (right picture) Dsh dissociates and can block the degradation of β-catenin. Subsequently β-catenin can accumulate and shuttles to the nucleus where it activates gene expression in concert with TCF/LEF (grey) transcription factors.
(Abbreviations: APC = adenomatosis polyposis coli; Dsh = Dishevelled; Frz = Frizzled; GSK3 = glycogen synthetase kinase 3; LRP = LCA-related phosphatise; TCF/LEF = T-cell factor/lymphoid enhancer factor; Wnt = Wingless Int)

specifically eliminate key factors of the pathways of interest in distinct cell populations.

Two transcription factors that are indispensable for early as well as later endoderm development, the forkhead box transcription factor a2 (Foxa2) and the SRY-related HMG box transcription factor 17 (Sox17; Ang and Rossant, 1994; Weinstein et al., 1994; Kanai-Azuma et al., 2002), have not been studied completely regarding their function and role. However, *Foxa2* and *Sox17* are among the first markers that can be detected in the endoderm (Lai et al., 1991; Kanai-Azuma et al., 1996). Fate mapping studies of different parts of the early DE reveal populations of cells that eventually express *Foxa2* or *Sox17* (Lawson et al., 1991; Tam et al., 2007). Genetic fate mapping marking *Foxa2* and *Sox17* positive cells exclusively would be the approach of choice to precisely identify derivatives.

To address the question of which pathways are important for the determination and the pattern of the endoderm and what exactly the fate of single cell populations is, mouse lines expressing Cre (causes recombination) under the control of the earliest makers known are the preferable tools. These lines would make it possible to analyze conditional knock outs of key factors within well-studied signalling pathways in a cell-type restrictive manner and to fate map the expressing cell populations by crossing these mice to reporter lines more precisely.

Factors of interest in this respect are *Foxa2* and *Sox17*, as mentioned above. The next section will therefore focus on the history, the expression of the two transcription factors and their integration into the transcriptional network regulating endoderm formation.

2.3. Important transcription factors during endoderm development: *Foxa2*

Foxa2 (forkhead box transcription factor a2), formerly known as *HNF3β* (hepatocyte nuclear factor 3 beta), was first identified in 1991 by its ability to bind to regulatory elements of liver-specific genes (Lai et al., 1991).

Foxa2 belongs to the family of Fox transcription factors that share a conserved DNA binding domain, the forkhead box. The protein family was given a unified nomenclature in 2000, including the subgroups FoxA-FoxS, based on sequence conservations (Kaestner et. al., 2000). Members of this family all belong to the helix-turn-helix class of proteins. Other FoxA subgroup members that show high similarity to *Foxa2* are *Foxa1* and *Foxa3*, formerly known as *HNF3α* and *HNF3γ*, respectively. The common DNA binding domain can also be found in their homolog: the developmentally important homeotic *fork head* gene of *Drosophila melanogaster* (Weigel and Jaeckle, 1990; Weigel et al., 1989; Lai et al., 1991). *Fork head* of *D. melanogaster* is required to correctly form terminal embryonic structures that contribute to tissues like the pancreas, the fore- and hindgut (Hartenstein et al., 1985; Jürgens and Weigel, 1988, Weigel et al., 1989). It is therefore not surprising that the mammalian homologues are expressed at the onset of DE development giving rise to different tissues such as the lung, the pancreas, the liver and the gut (Ang et al., 1993, Monaghan et al., 1993, Sasaki and Hogan, 1993). The forkhead box gene *Foxa2* is expressed throughout development persisting into adulthood (Cockell et al., 1995; Wu et al., 1997), starting at E6.0 in the VE at the anterior side of the embryo and a few hours later at E6.5 in the embryonic epiblast at the posterior side where gastrulation occurs (Ang et al., 1994). Foxa2 is the first member of the Foxa subgroup to be expressed, followed by Foxa1 and then Foxa3 (Ang et al., 1993; Monaghan et al., 1993).

The expression of *Foxa2* in the AVE, where it directly activates *Otx2*, points to its role in the establishment of the anterior-posterior axis (Kimura-Yoshida et al., 2007). *Otx2* expression in the AVE is required for fore- and midbrain induction (Rhinn et al., 1998). Moreover, *Foxa2* is essential for the expression of *Dkk1* (Glinka et al., 1998), *Cerl* (Piccolo et al., 1999) and *Lefty* (Perea-Gomez et al., 2002) three important Wnt and Nodal antagonists in the visceral endoderm at pre- to early streak stages (Kimura-Yoshida et al., 2007). Expression of *Foxa2* during gastrulation in the extra-embryonic VE has been shown to be essential for the elongation of the primitive streak and in the embryonic tissues for the formation of the organizer using tetraploid aggregations (Dufort et al., 1998). Using this technique one takes advantage of the fact that in aggregations of tetraploid cells (obtained by fusing 2-cell stage embryos) and ES cells, tetraploid cells can only give rise to extraembryonic tissues while the ES cells form the embryo proper (Nagy et al., 1990; Kaufman et al., 1990). This allows for the discrimination between functions of a single gene in extra-embryonic versus embryonic tissues: phenotypes with an origin in extra-embryonic tissues that appear earlier and do not allow for further development can be rescued and it becomes possible to focus on the phenotype with embryonic origins.

The expression of *Foxa2* in the epiblast at this stage overlaps partially with the expression of *Goosecoid*, a mediator of RA signalling (*Gsc*; Blum et al., 1992), and *Chordin*, a secreted bone morphogenetic protein (BMP) antagonist (*Chrd*; Piccolo et al., 1996). More precisely, it overlaps at the anterior tip of the primitive streak at mid-streak stages and in the node at the late streak stage, always coinciding with the population of cells having organizer function (Kinder et al., 2001).

Foxa2, compensatory with *Foxa1*, is also important for the branching morphogenesis of the lung during embryogenesis (Wan et al., 2005) while absence of *Foxa2* expression in the lung leads to failure of transition to air breathing at birth due to the fact that it is necessary for the activation of genes mediating surfactant protein

Introduction

production and lipid synthesis (Wan et al., 2004). The initiation of liver development is also compensatory depending on *Foxa1* and *Foxa2* expression (Lee et al., 2005a). Pancreatic α-cell differentiation, however, relies on *Foxa2* expression solely, although specification of the precursors is not affected (Lee et al., 2005b).

Later, in the adult mouse, *Foxa2* is mainly expressed in lung, liver (Lai et al., 1991) and in the pancreas (Lantz et al., 2004; Sund et al., 2001). In the lung it marks type II pneumocytes, regarded as the "stem cells" of the lung (Mason et al., 1997), whereas in liver and pancreas it plays a role in hepatic and pancreatic metabolism respectively (Lee et al., 2002; Zhang et al., 2005; Lantz et al., 2004; Sund et al., 2001; Kaestner et al., 1999; reviewed by Friedman and Kaestner, 2006). The function of *Foxa2* in Parkinson's disease has yet to be fully understood; however, *Foxa2* is essential for the generation of dopaminergic neurons in embryogenesis, and has also been associated with the progressive loss of these neurons in the adult organism (Kittappa et al., 2007).

It was shown that *Foxa2* expression is indispensible for the development and formation of the embryonic gut as well as for the axial mesoderm (Ang and Rossant, 1994; Weinstein et al., 1994). Mice lacking *Foxa2* expression die around E9-E10 due to the fact that they develop neither a notochord nor a distinct node (Ang and Rossant, 1994; Weinstein, 1994). The absence of a correctly developed organizer leads to secondary defects in the organisation of the somites and the neural tube as well as defects in the fore- and midgut (Ang and Rossant, 1994; Weinstein et al., 1994). In contrast, *Foxa1*$^{-/-}$ mutants die due to abnormal glucagon secretion and hypoglycemia about 10 days after birth (Kaestner et al., 1999). *Foxa3*$^{-/-}$ mutants, however, only show defects in glucose homeostasis and associated gene expression during prolonged fasting (Shen et al., 2001). The fact that consensus DNA binding sequences for all three transcription factors, FOXA1, 2 and 3, overlap suggests that their molecular function may be redundant (Lai et al., 1991).

Taken together, *Foxa2* has a broad range of activity. In embryonic development, *Foxa2* is expressed not only in tissues derived from of the endoderm but also in all cell populations with organizer potential and its derivatives. This includes extra-embryonic tissues such as the AVE which have organizer function (Rhinn et al., 1998; Kinder et al., 2001).

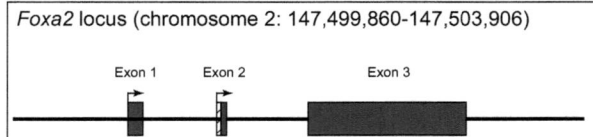

Figure 9: The genomic locus of *Foxa2*
The *Foxa2* gene consists of three exons (dark grey boxes). The ORF starts within exon 2 and ends in exon 3. The gene is located on chromosome 2.
(Abbreviations: ORF = open reading frame)

2.4. Important transcription factors during endoderm development: *Sox17*

Sox17 (<u>s</u>ex determining region Y related HMG (<u>h</u>igh <u>m</u>obility <u>g</u>roup) <u>box</u> transcription factor <u>17</u>), the other transcription factor known to be expressed early during endoderm formation, was first identified in a mouse testis cDNA screen in 1996 (Kanai et al., 1996).

Sox17 belongs to the diverse group of Sry-related HMG-box transcription factors that contain the HMG box, first identified in the sex determination gene, *Sry* (Nagamine et al., 1987). Approximately 33 *Sox* genes have been currently identified and they are subdivided in 10 different groups, namely A to J, according to their sequence conservation in the HMG box and other structural motifs on the one hand and according to functional studies on the other hand (Bowles et al., 2000). All members of this group of transcription factors are broadly involved in early embryonic development, regulating different aspects of determination, not only in testis development but also cardiovascular development, neural development, B cell and endoderm development (for review see Lefebvre et al., 2007).

Sox17 belongs to subgroup F, along with *Sox7* and *Sox18*, and is known to be first expressed in the visceral endoderm from E6.0 on (Kanai et al., 1996). From E7.0 on it can also be detected in the definitive endoderm, more precisely the ADE and in the embryonic hindgut. After E9.5, *Sox17* expression could no longer be detected in the endoderm but was instead observed in vascular endothelial cells of the dorsal aortae and intersomitic blood vessels (Matsui et al., 2006).

In the context of Wnt-signalling *Sox17* binds to β-catenin thereby inhibiting transcription of β-catenin target genes (Takash et al., 2001). In contrast several transcriptional targets, including *Gata4* (*GATA binding protein 4*), *Foxa1* and *Foxa2*, playing roles in the development of the early endoderm were identified in *Xenopus* (Sinner et al., 2004).

It was shown that *Sox17*, redundantly with *Sox18*, has a pivotal role in the establishment of the early vasculature in embryonic development and that it affects cardiac development as well as postnatal angiogenesis (Matsui et al., 2006; Sakamoto et al., 2007). More recently, Kim et al. also reported the expression of *Sox17* in fetal haematopoietic stem cells and its requirement for the maintenance of fetal and neonatal but not adult haematopoietic stem cells (Kim et al., 2007).

Targeted deletion of the *Sox17* gene in mice, however, results in embryonic lethality around E9.5; embryos are unable to turn, a morphogenetic process depending on proper endoderm development. *Sox17* mutants show a reduced foregut, in addition to mid- and hindgut defects (Kanai-Azuma et al., 2002). All these observations point to the essential role of *Sox17* in the formation and establishment of the endoderm prior to vasculogenesis.

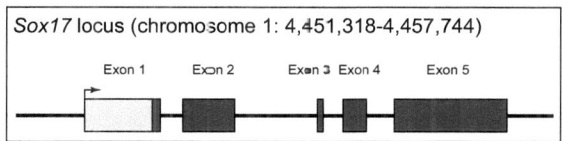

Figure 10: The genomic locus of *Sox17*
The *Sox17* gene consists of 5 exons (dark grey boxes; yellow box = newly elongated exon 1). The ORF starts within exon 4 and ends in exon 5. It is located on chromosome 1.
(Abbreviations: ORF = open reading frame)

2.5. Stem cells – a model for embryonic development and an indispensible tool to analyze gene function

Embryogenesis and stem cell research are closely related topics. Factors important for embryonic development of certain lineages are shown to be important for *in vitro* differentiation of ES cells and vice versa (e.g. retinoic acid; for review see Keller, 2005). *In vitro* differentiation protocols were established on the basis of knowledge gained by developmental studies *in vivo*, e.g. human ES cells were differentiated into endoderm derivatives (D'Amour et al., 2006, Kroon et al., 2008).

Research on both sides can therefore profit from one another and especially early expressed genes marking basic lineage decisions, like *Foxa2* and *Sox17*, are of interest in *in vitro* differentiation systems. The next part will therefore concentrate on ES cells and *in vitro* differentiation systems.

Stem cells are divided in two major groups: embryonic and adult stem cells.

In general all stem cells share certain properties by which they are defined as stem cells. Firstly stem cells theoretically have the potential to divide infinitely. Secondly while dividing they reproduce, known as self-maintenance, and thirdly they are able to differentiate into at least one other more differentiated cell. This cell either might be a tissue specific precursor cell, still dividing and differentiating, or a terminally differentiated cell.

ES cells are generated from the ICM of blastula stage embryos and can give rise to all different tissues derived from the three principle germ layers *in vivo* and *in vitro*, respectively (Keller, 1995; Rohwedel et al., 1994; Brustle et al., 1999; Rossant, 2001; Smith 2001). Therefore they are considered pluripotent. Only cells from the zygote up to the 2-cell stage (and with restrictions the 4-cell stage) of a mouse embryo can be called totipotent, because a single cell could theoretically build the whole organism (for review see Torres-Padilla, 2008).

The first murine ES cell line was isolated in 1981 (Martin, 1981), followed by the first rhesus and human ES cell lines (Thomson et al., 1995; Thomson et al., 1998). Using ES cells that can be kept in an undifferentiated state *in vitro* it became relatively easy to manipulate the mouse genome and establish multiple mouse lines carrying genomic alterations as models of human genetic diseases (for review see Bedell et al., 1997). Many pathways and factors indispensable for keeping ES cells in culture are still not identified. ES cells from different organisms seem to require different factors to maintain their pluripotency. Mouse ES cells, for instance, need the presence

Introduction

of a pleiotropic cytokine, leukaemia inhibitory factor (LIF; Smith, 1991), while human ES cells are cultured using BMP4, an inhibitor of the MAP (mitogen-activated protein) kinase transduction pathway (Qi et al., 2004). LIF serves as a factor for the preservation of pluripotency and self-renewal via the STAT-3 pathway (Heinrich et al., 1998; Niwa et al., 1998; Constantinescu, 2003). *Oct4* (*Octamer-binding transcription factor 4*), *Sox2* and *Nanog* (Nanog homeobox) are key regulators in the formation and/or maintenance of the epiblast of the ICM and are also necessary for self-renewal and maintenance of pluripotency in ES cells (Nichols et al., 1998; Niwa et al., 2000; Avilion et al., 2003; Mitsui et al., 2003).

Adult stem cells are tissue-specific precursor cells found in the adult organism with a restricted differentiation potential (for review see Raff, 2003; Zuk et al., 2002; Alvarez-Buylla et al., 2002). The best investigated representative belonging to this type of so-called multipotent stem cells is the adult haematopoietic stem cell, resting in the bone marrow (Reya et al., 2001).

Other types of stem cells are embryonic germ (EG) cells and embryonic carcinoma (EC) cells. EG cells are derived from the genital ridge of the embryo from pre- and post-migratory as well as from migratory germ cells (Resnick et al., 1992; Matsui et al., 1992; Shamblott et al., 1998; Durcova-Hills et al., 2001). They, *in vivo*, differentiate into female and male germ cells. *In vitro* they are remarkably similar to mouse ES cells lines but have a limited proliferation potential (Donovan and de Miguel, 2003; Donovan and Gearhart, 2001; Labosky et al., 1994). EC cells were first identified from a cancer called teratocarcinoma. These cells are able to form teratoma, carcinoma, consisting of several differentiated tissues. Their unlimited ability to self-renew *in vitro* and their great similarity with cells of the ICM of the blastocyst cleared the way for the culture of ES cells (Kleinsmith and Pierce, 1964).

A more recently generated form of stem cell in human and mouse is the so-called induced pluripotent stem (iPS) cell (Takahashi et al., 2007; Takahashi and Yamanaka, 2006; Nakagawa et al., 2008). By retroviral overexpression of four important factors that are usually found highly expressed in ES cells, the scientists obtained dedifferentiated fibroblast cells with ES cell-like character regarding their expression profile, their culture, their morphology and their potential to differentiate into different lineages derived from the three germ layers, including the germ line (Takahashi and Yamanaka, 2006; Takahashi et al., 2007; Okita et al., 2007). The four factors used for the dedifferentiation are *Oct4*, *Sox2*, *Kif4* (kinesin family member 4), and *c-Myc* (cellular myelocytomatosis oncogene; Takahashi and Yamanaka, 2006). The subsequent finding that *c-Myc* overexpression is not necessarily needed for this process (Nakagawa et al., 2008) is a clear advancement in light of *c-Myc*s protooncogenic potential development (Okita et al., 2007; Vennstrom et al., 1982). In another study adult human somatic cells were reprogrammed to a pluripotent state using *Oct4*, *Sox2*, *Nanog*, and *Lin28* (lin-28 homolog; Yu et al., 2007). Most recently, Takashi et al. succeeded in dedifferentiating adult mouse hepatocytes and gastric epithelia cells into pluripotent cells (Aoi et al., 2008).

ES and iPS cells have a great potential in cell replacement therapies because in principle any amount and any cell type could be produced in *in vitro* differentiation systems.

To be able to differentiate different cell types from ES cells *in vitro* directed and more efficiently, a deep knowledge about the embryonic development and the single steps of differentiation leading to certain cell types is necessary. Indeed, several lines of evidence suggest that *in vitro* differentiations (IVD) proceed analogously to the *in vivo* development allowing the translation of the *in vivo* knowledge to directed *in vitro* differentiation assays. Up to now there are different examples of successfully *in vitro* differentiated ES cells, e.g. pancreatic hormone-expressing endocrine cells or glucose-responsive insulin-secreting cells from human ES cells *in vivo* supporting this hypothesis (D'Amour et al., 2006; Kroon et al., 2008). But as mentioned before, there are still many pathways which parts have to be unravelled regarding the development into different lineages. Also micro RNAs (miRNAs) e.g., are known to play important roles in early development and are capable of inhibiting translation, and consequently downregulating protein levels, very fast and efficiently. Thus miRNA target analysis should also be paid attention to.

In order to identify and isolate cells at a certain developmental stage to be able to investigate involved signalling pathways and molecules marking ES cells genetically with fluorescent proteins is an important first step. This isolation and subsequent separation might help investigating single steps in *in vitro* differentiations of pure populations. The results of this analysis could help to optimize the differentiation of certain lineages *in vitro* in regards to usage of *in vitro* differentiations for cell replacement therapies. ES cells, like EC cells, are capable of forming teratoma when not fully differentiated and injected into the organism (Iles, 1977; Choi *et al.* 2002; Wakitani *et al.*, 2003; Nussbaum *et al.*, 2007).

Secondly, genetic marking is a necessary tool to optimize cell culture conditions and study the impact of diverse factors more easily. These techniques also allow for the use of high-throughput assays compared to time-consuming and indirect molecular characterization via RT-PCR for example (Tada *et al.*, 2005). Furthermore, analyses with cellular resolution can be carried out that would not be possible in older methods which are only able to analyze a population as a whole. Factors known for marking certain lineages could be fused with fluorescent proteins then used as indicators for miRNA activity and impact in miRNA inhibition assays.

Thirdly, the more fundamental problem of analyzing gene functions *in vivo* also depends on genetic alterations of ES cells that still have the ability to give rise to all cells of the organism.

The different approaches for the genetic marking and the mutation of ES cells and the generation of whole organisms carrying those alterations are discussed further in the next chapter.

2.6. Alterations of the genome

Alterations of the genome to analyze gene function might become necessary at any step of a project, especially as the genomes of various species are sequenced (Lander *et al.*, 2001; Waterston *et al.*, 2002; Adams *et al.*, 2000) and much genomic data lacks functional annotation. Mutations are therefore widely used in *in vitro* and *in vivo* assays to assess gene functions.

Generally there are many different approaches for mutations of a cell or a whole organism. In principle they can be subdivided into random versus targeted approaches. Random mutations can be generated by irradiation or treatment with chemicals, e.g.; in this case the specific mutation or mutations are unknown with regard to the affected gene, or the exact kind of alteration. It should also be mentioned that these mutational alterations were often done with whole organisms achieving chimeras with many differently modified cells and only those mutations that affected germ cells and were transferred to the next generation could be analyzed.

Random alterations also incorporate those generated by random integration of a transgene – regardless if the transgene is a simple bacterial plasmid or a <u>b</u>acterial <u>a</u>rtificial <u>c</u>hromosome (BAC) although the latter method is believed to reflect endogenous expression more precisely because of its size theoretically all *cis*-regulatory elements should be found on a BAC. Cell or mouse lines generated this way can in principle allow for the identification of the integration site (see also the localization of gene trap insertion sites: Gossler *et al.*, 1989; Wurst *et al.*, 1995), though this approach opens the possibility of interrupting <u>o</u>pen <u>r</u>eading <u>f</u>rames (ORF) of important genes or any other regulative sequence, an incidence that is taken advantage of by the gene trap method (Gossler *et al.*, 1989). The expression of the transgene, however, can differ in tissues due to the integration sites within active or inactive DNA. This is especially true for overexpression using strong eukaryotic promoters driving transgenes where toxicity might be a serious issue.

Genetic targeting of ES cells was first performed in 1987 when Mario Capecchi and Kirk Thomas selectively inactivated the *Hprt* (*hypoxanthine-guanine phosphoribosyl-transferase*) gene in the mouse (Thomas and Capecchi, 1987). This method became a standard technique during recent years and resulted in the 2007 Nobel Prize for medicine being awarded to Mario Capecchi for the first proof of homologous recombination in mammalian cells (Thomas and Capecchi, 1986), Oliver Smithies for homologous recombination in human cells with circular DNA at the same time (published a few month earlier: Smithies *et al.*, 1985) and Sir Martin Evans, the discoverer of embryonic stem cells (Evans and Kaufman, 1981), for the first injection experiments of embryonic carcinoma cells into blastocysts and the generation of chimaeras (Papaioannou *et al.*, 1975). In genetic targeting, the *gene of interest* is modified in ES cells via homologous recombination. As in the case for random integrations, these cells can be kept in an undifferentiated state *in vitro* and can subsequently be injected into donor blastocysts that are transferred into the receptive uterus of a pseudo-pregnant foster mouse. The resulting chimaeras consisting of gene targeted cells (originating from the gene targeted the ES cells) and wild type cells (derivatives of the donor blastocyst cells) are backcrossed to wild type mice. If the germ cells of the chimaeras contain gene targeted cells the alteration can be transmitted through the germline. Thus, mice carrying the alterations of the gene of interest in all cells of the body can be produced. The advantage of this procedure is that the genomic alteration is defined and that expression of a transgene can be positioned under the control of the natural promoter and all other activating or repressing cis-regulatory elements. Expression of fluorescent proteins or β-galactosidase under the specific control of any gene allows one to easily and most accurately follow or report its expression. Nevertheless, this approach has several limitations. Heterozygous embryonic lethality (VEGF: vascular endothelia growth factor; Carmeliet et al., 1996; Ferrara et al., 1996) and compensation effects by redundant genes (Matsui et al., 2006; Sakamoto et al., 2007) might occur and do not allow for the analysis of gene function as easily as planned.

With the establishment of the technology of site specific Cre (Sternberg and Hamilton, 1981) loxP (locus of X over in P1; Hoess et al., 1982) recombination in vitro (Sauer and Henderson, 1988) and in vivo (Orban et al., 1992; Lakso et al. 1992) it became possible to conditionally delete genes in different cell types or to induce deletion at different time points (for review see Rajewsky et al., 1996; Kühn and Schwenk, 1997). Cre, a member of the λ-integrase family found in the bacteriophage P1, recombines loxP sites, 34bp long target sites, and in doing so it cuts out the DNA-sequence flanked by sites orientated in the same direction (Sauer et al., 1988; Gu et al., 1993) or inverts the DNA if the orientation of the loxP sites is opposing (Kano et al., 1998; Lam and Rajewsky, 1998). Cre recombination activity does not depend on any cofactors (Abremski and Hoess, 1984; Hoess et al., 1990).

Tracing cells for their fate over the time of development by conditional activation of marker genes like β-galactosidase or any fluorescent protein is only one possibility. LoxP-flanked stop-codons in front of the open reading frames (ORF) of any gene could be used to switch on genes this way. The genetic cell lineage labelling can be regarded as the more precise way to track cells compared to cell labelling with dyes (Tam et al., 2007).

A second approach that can be realized using this technique is the conditional deletion of one or more genes that are flanked by loxP sites by expression and recombination activity of Cre recombinase. As stated before, this can be either tissue- or time-specific by expressing Cre under the control of a certain promoter. Cre can also be fused to a hormone-inducible protein domain, e.g. to the estrogen receptor (ER), that allows one to activate Cre by hormone-induced translocation of the protein from the cytoplasm to the nucleus (Metzger et al., 1995). Using a mutated version of the estrogen receptor, ERT2 (estrogen receptor type 2), in combination with exogenous tamoxifen as the inducer one avoids cross-activation of the Cre by an organism's own hormones (Feil et al., 1996; Feil et al., 1997). Timely induction, especially during embryonic development, and leakiness can complicate the handling of this system.

Deletion of genes using Cre recombinase is one approach to analyze gene function in vivo. Because the alteration takes place on the genomic level it totally abolishes the protein production and therefore does not perfectly mimic the situation in most mutations just resulting in partial protein dysfunction. Usually the natural effect of mutations can be mimicked by down regulating the protein expression using RNA interference (RNAi; Lewis et al., 2002; McCaffrey et al., 2002; Sorensen et al., 2003; Brummelkamp et al., 2002; Lee et al., 2002; McManus et al., 2002; Miyagishi and Taira, 2002; Paddison et al., 2002; Paul et al., 2002; Sui et al., 2002; Yu et al., 2002; Kawasaki and Taira, 2003). SiRNA (small interfering RNA) models can be used to study mutations on a more natural state, but one should be aware of possible pleiotropic effects. Exerting influence on differentiation processes of the natural miRNAs (micro RNAs), knock-downs for specific miRNAs or transgenics over-expressing miRNAs could be a tool used in in vitro experiments. Later on the obtained knowledge could also be transferred by transient expression of an organism's own miRNAs to prevent or to force certain lineage decisions.

2.7.1. MiRNAs in development and differentiation

MiRNAs are small non-coding RNAs, approximately 21 nucleotides long and have been shown to regulate gene expression. They impact signalling networks by regulating alternative splicing, inhibiting mRNA translation or forcing mRNA degradation (Makeyev et al., 2007; Lee et al., 1993; Olsen and Ambros, 1999; Dugas and Bartel, 2004). MiRNAs are highly conserved compared to endogenous small interfering RNAs (siRNAs; Ambros et al.,

2003; for review see Bartel, 2004) and might be evolutionarily older than first assumed based on the recent identification of miRNAs in the unicellular algae *Chlamydomonas reinhardtii* (Molnar et al., 2007; Zhao et al., 2007).

MiRNAs are known to play important roles in cancer, maintenance of mature cell function, regeneration, embryonic development and stem cell self-renewal, maintenance and differentiation (Houbaviy et al., 2003; Suh et al., 2004; Bernstein et al., 2003; Wienholds et al., 2003; Ota et al., 2004; Calin et al., 2004; He and Hannon, 2004; Joglekar et al., 2007; for review see Hwang and Mendell, 2006; Stefani and Slack, 2008).

In *Xenopus laevis* recent studies show that miRNAs play a role in early development through the inhibition of Nodal signalling by reducing the expression of one of its receptors (Martello et al., 2007). In addition, the blockage of the two miRNAs miR15 and miR16 could rescue dorsal mesoderm induction in embryos with suppressed Wnt/β-catenin signalling. The observation that gradients of the miRNAs and Wnt/β-catenin are reciprocal suggests that suppression of miRNAs might be one mechanism by which Wnt promotes Nodal-signalling. Interestingly, the targeting sites for miRNAs regulating the Nodal receptor are conserved in mammals, but not in zebrafish (Martello et al., 2007). Indeed the Nodal signalling pathway is also controlled by miRNA-mediated modulation in zebrafish. MiR430 directly decreases the expression of *Squint*, a member of the Nodal family, and interestingly also regulates *Lefty*, an antagonist of Nodal (Choi et al., 2007). MiR430 seems to be relevant for the fine-tuning of Nodal activity by influencing agonist and antagonists, further supported by the fact that overexpression of either *Squint* or *lefty* does only result in a phenotype and that the mutation of miR430 complementary sites leads to a disruption of early development (Choi et al., 2007).

In addition to their role in early development which has yet to be shown in mammals, distinct miRNAs have other functions in later development of several organs. In the mouse brain miR124 regulates the switch to neuronal differentiation by inhibition of a regulator of alternative splicing (PTB = polypyrimidine tract binding protein; Makeyev et al., 2007). MiR1, e.g., a miRNA highly expressed in skeletal muscle and heart evolutionary conserved across different species from fly to human (Lagos-Quintana et al., 2002; Kwon et al., 2005; Sokol et al., 2005) was shown to play in important role in the proper establishment and maintenance of cardiac tissue (Zhao et al., 2005, Zhao et al., 2007). In *in vitro* assays miR1 was shown to regulate differentiation versus proliferation with high and low levels, respectively (Chen et al., 2006), in line with *in vivo* overexpression which resulted in a thinner ventricular wall in the mouse (Zhao et al., 2005). More recent work has demonstrated that selective ablation of miR1-2 (second copy of miR1) leads to hyperplasia and consequently to prenatal or early postnatal death (Zhao et al., 2007). Increased miR1 levels are also observed in and associated with heart ischemia animal models and human coronary artery disease (Yang et al., 2007). Interestingly, the two copies of miR1 (miR1-1 and miR1-2) are expressed in a similar, but not identical, spatial and temporal pattern and, as shown by the distinct phenotype caused by removal of miR1-2 alone, do not have perfectly redundant actions. MiR133 is transcribed along with both miR1-1 and miR1-2 copies but in contrast to miR1 it increases proliferation and decreases myocyte differentiation (Chen et al., 2006). As in the case for miR1, its reduction is associated with human cardiac hypertrophy (Sayed et al., 2007, Care et al., 2007). These data also suggests a role in fine-tuning of the two miRNAs during muscle differentiation, but also show a cooperative function in repression of heart hypertrophy.

Upregulation of another miRNA, miR155, is associated with different lymphomas, breast cancer and a poor prognosis in lung cancer (Eis et al., 2005; van den Berg et al., 2003; Kluiver et al., 2005; Iorio et al., 2005; Yanaihara et al., 2006). Altered distributions of different types of immune cells show its complex role in the adaptive immune response (Rodriguez et al., 2007). Interestingly, miR155 is part of the *BIC* (*B-cell integration cluster*) gene, originally identified as a site of frequent integration of an avian virus leading to B-cell lymphoma induction (Clurman et al., 1989). BIC itself codes for a 1700 nucleotide long polyadenylated RNA that is spliced but lacks an ORF. Except for the 100 nucleotide long sequence that encodes for miR155 (precursor) the *BIC* gene is poorly conserved, suggesting that the major role of *BIC* is miR155 expression (Tam et al., 1997).

In addition to their roles during development and differentiation miRNAs are very likely regulators of stem cell self-renewal, since loss of Dicer, an important actor of the miRNA pathway (discussed in the following section) leads to the loss of stem cell populations (Bernstein et al., 2003; Wienholds et al., 2003). Certain miRNAs – some of which are evolutionary conserved between mouse and human – are specifically expressed in pluripotent ES

cells and not in differentiated cells or adult tissues, thus also pointing to a role of miRNAs in stem cell self-renewal (Houbaviy et al., 2003; Suh et al., 2004).
Fast downregulation of stem cell maintenance and activation of lineage-specific genes is necessary for stem cell differentiation. MiRNAs are most likely involved in this rapid switch of gene expression since Dicer1 knock-out ES cells that lack mature miRNAs fail to differentiate into ectoderm, mesoderm or endoderm (Kanellopoulou et al., 2005). Later in vitro differentiation is also influenced by miRNAs, shown for miR181 where ectopic expression causes an increase of the B-cell fraction in vivo and in vitro (Chen et al., 2004).

2.7.2. The miRNA pathway: from transcription to target interaction

MiRNAs were first identified in C. elegans as short double stranded RNAs that repress their targets translationally by binding to specific sites in the 3' UTR of the mRNA (Lee et al., 1993; Olsen and Ambros, 1999). Originally shown for for siRNA in RNAi experiments (Elbashir et al., 2001), perfectly basepairing of miRNAs, as it is common in plants, leads to cleavage of the target RNA miRNA duplex between position 10 and 11 (Dugas and Bartel, 2004).MiRNAs are coded in the genome and are transcribed by polymerase type II into RNA as long precursors of the mature miRNA, the so-called pri-miRNAs (see figure 11).

Pri-miRNAs have a complex secondary structure and parts of this structure that carry the miRNA, the hairpins, are recognized by Drosha, a RNAse III that cleaves pri-miRNA into the shorter, ~60-70bp long single hairpins, known as pre-miRNA.

Pre-miRNAs interact with Exportin 5 (exp5), a protein that triggers the nuclear export of miRNAs in cooperation with its co-factor Ran GTP dependently (Yi et al., 2003; Bohnsack et al., 2004; Lund et al., 2004; Lei and Silver, 2002). Cullen et al. showed that the loop seems to have only little influence on the export to the cytoplasm but that binding to of the double-stranded pri-miRNA is supported by 2nt 3' overhang, compared to blunt ends, and inhibited by 5' overhangs (Cullen et al., 2004).

Several observations suggest that exp5 might be the bottle-neck within the RNAi pathway (Yi et al., 2005; Grimm et al., 2006). In the cytoplasm, exp5 releases the pre-miRNA followig the hydrolysis of Ran-GTP to –GDP. Subsequently, Dicer binds the pre-miRNA initially recognizing it by the characteristic 2nt 3' overhang. Dicer then aligns itself along the RNA double strand and leaves dsRNAs with a ~19-23bp stem, a 5' mono-phosphate and two 2nt 3' overhangs. After Dicer cleavage the miRNA duplex is incorporated into the RNA-induced silencing complex (RISC; Hammond et al., 2000). The first component of the RISC that could be identified is Argonoute2 (Ago2; Hammond et al., 2001; Martinez et al., 2002). Ago2 is the catalytic compartment of RISC that is responsible for target cleavage. It could also be shown that all Ago proteins known so far co-immunoprecipitate with Dicer, suggesting that Ago binding to the miRNA duplex is initialized during binding of Dicer (Sasaki et al., 2003). The RISC was shown to favour the strand of the RNA duplex with less stability at the 5' end for final incorporation, although both strands are initially incorporated into RISC (Matranga et al., 2005; Rand et al., 2005; Leuschner et al., 2006). This strand will then serve as the miRNA.

Data suggest that siRNA and miRNA at this point are completely interchangeable and that the following action of RISC solely depends on the level complementarity regarding the target mRNA (Hutvagner and Zamore, 2002; Doench et al., 2003; Zeng et al., 2003).

Target cleavage one possible RISC action. The guide strand (siRNA strand within the RISC) complementary binds to the sites of the target mRNA which is cleaved by the endonuclease activity of Ago2. With high specificity the target is cleaved between the 10th and 11th nucleotide pairing to the guide strand, counting from the 5' end of a 21mer siRNA (Rand et al., 2005; Leuschner et al., 2006).The guide strand thereby remains intact and can serve as target recognition site for multiple rounds. Released fragments of the target mRNA are presumably degraded by cellular exonucleases.

Translational repression is the other possibility of RISC action. MiRNAs show only partial complementarity towards their target mRNAs which for this reason are hard to predict (Brennecke et al., 2005).

The base-pairing at the 5' proximal "seed" region (position 2-8) of the miRNA to the mRNA target sequence first seemed especially crucial for recognition (Lewis et al., 2005; Brennecke et al., 2005). Later studies could show that base pairing in the seed region is not as important as first assumed (Didiano and Hobert, 2006). MiRNAs were shown to cause translational inhibition of their targets in *C. elegans* (Lee *et al.*, 1993; Wightman *et al.*, 1993), *D. melanogaster* (Brennecke *et al.*, 2003) and human (Zeng *et al.*, 2003), a mechanism, which is far from being understood. A possible explanation for this phenomenon could be inhibition of translational initiation because targeted mRNAs are less often associated with ribosomes (Pillai *et al.*, 2005; Moss *et al.*, 1997). Although miRNAs were thought to only inhibit translation of the target mRNA more recent data suggest that RISC also causes decrease of mRNA levels by cleavage (Bagga *et al.*, 2005; Jing *et al.*, 2005; Lim *et al.*, 2005; Liu *et al.*, 2005; Rehwinkel *et al.*, 2006; Wu *et al.*, 2006).

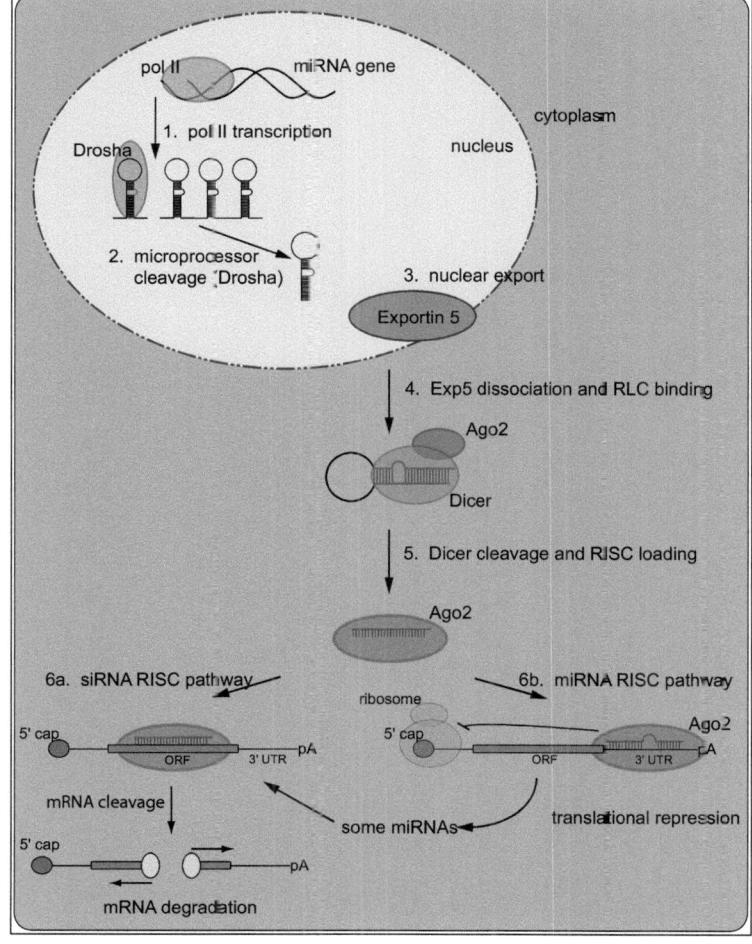

Figure 11:
The miRNA pathway

MiRNA genes are usually transcribed by a polymerase type II (pol II; 1.).
The pri-miRNA is then cleaved into hairpin like molecules of RNA with approximately 60-70nt length by Drosha (2.). The so-called pre-miRNA is exported to the cytoplasm by Exp5 (3.) where it dissociates again and is bound to the RLC (4.). The miRNA is then cleaved by Dicer (5.). Subsequently the guide miRNA strand out of the duplex is incorporated into RISC and can now act as an siRNA leading to mRNA cleavage (6.a) or as an miRNA by binding to the 3'-UTR of the target mRNA and blocking its translation (6.b).
(Abbreviations: Ago = Argonaute; Exp5 = Exportin-5; miRNA = micro RNA; nt = nucleotides; ORF = open reading frame; pA = polyadenylation; pol II = polymerase type II; RISC = RNA induced silencing complex; RLC = RISC loading complex; siRNA = small interfering RNA; UTR = untranslated region)

3. Rationale

Part I – Endoderm differentiation *in vivo*

The signals and signal transduction pathways that are crucial for the first important differentiation processes in the endoderm have yet to be fully understood. Two transcription factors, *Foxa2* and *Sox17*, are known to play essential roles in early endoderm development but they have not been studied completely regarding their function. *Foxa2* as well as *Sox17* are among the first markers that can be detected in the endoderm (Kanai et al., 1996; Ang and Rossant, 1994; Weinstein et al., 1994) but it remains unclear as to what organs, tissues and cell types *Foxa2*- and/or *Sox17*-positive progenitor cells give rise to. Identifying their progeny, unravelling their interactions and integrating them more accurately into the transcriptional network would help and support the progress and understanding of *in vitro* differentiation systems for stem cells and therefore could eventually result in therapeutic applications. In this context the analysis of intrinsic and extrinsic signals and pathways that regulate endoderm differentiation is also an important step. The aim of this thesis was to establish mouse lines expressing Cre recombinase under the transcriptional control of all *Foxa2* and *Sox17* regulatory elements to trace the lineage of cells in the mouse embryo that express these two earliest genes known to mark the endoderm, in order to answer the question of what organs these cells will give rise to.

In addition, the *Foxa2*$^{iCre/+}$ mouse line was used to investigate the specific role of Wnt signalling in the differentiation of cells expressing *Foxa2*^{iCre} by conditional deletion of the *β-catenin* alleles in those cells with the help of a mouse line with conditional alleles for *β-catenin*.

Part II – Endoderm differentiation *in vitro*

The goal in stem cell research is to understand how stem cells keep the ability to self-renew, how they are maintained into adulthood and what signals are necessary to accomplish the generation of specific cell types in *in vitro* differentiations of pluripotent (embryonic) stem cells.

Non-coding RNAs (ncRNAs) play important roles in embryonic development by coordinating cell-fate (for review see Cheng et al., 2005) and in ES cells by regulating stem cell maintenance and asymmetric cell division (Hatfield et al., 2005) in a post-transcriptional manner. Micro RNAs (miRNAs), a class of ncRNAs, can inhibit mRNA from being translated into protein or can lead to mRNA degradation. They are involved in cell proliferation, cell death, carcinogenesis, self-renewal and differentiation (for review see Hwang and Mendell, 2006; Stefani and Slack, 2008).

It is important to identify targets of miRNAs to understand their individual roles in the regulation of transcriptional and signaling networks. Computational algorithms allow for the prediction of up to 300 target mRNAs per miRNA (Brennecke et al., 2005; Bartel, 2004). A systematic combination of both predictive and experimental approaches is necessary to identify, verify and classify miRNAs.

In order to identify miRNAs that regulate endoderm development an experimental test system for miRNAs was built up – thereby also obtaining a deeper understanding about the differentiation of ES cells into endoderm – another aim of this thesis was to establish an *in vitro* system for endoderm differentiation. In principle this would allow for the ability to compare *in vivo* and *in vitro* data and having an easily accessible model system for endoderm differentiation

With the help of ES cell clones that carry fluorescent fusions of the two transcription factors that have important functions in endoderm development (*Foxa2* and *Sox17*) the differentiation system was established and tested regarding its use as a miRNA test system. The fusion proteins and their endogenous counterparts of the transcription factors do not only share all transcriptional regulatory elements but also have the endogenous 3'-UTR as a common feature. Because mRNAs are targeted by miRNAs through binding of specific sequences within the 3'-UTR und because this binding subsequently leads to translational blockage, less protein is expected for both the endogenous protein and its fluorescent fusion, latter of which will result in lower fluorescence.

4. Results

For the analysis of the fate of specific subpopulations of cells in the embryonic endoderm, knock-ins of recombinases under the control of all transcriptional elements of a specifically expressed marker gene is the approach of choice in mice, because they reflect the endogenous expression of the gene as accurately as possible (see figure 12). This is due to the fact that the knock-in, here the recombinase, is in a genetically non-altered environment, compared to transgenic vectors or bacterial artificial chromosomes (BAC) based lines in which the recombinase is randomly integrated into the genome, potentially interrupting open reading frames of important genes. BACs should basically (compared to other plasmid vectors) carry all cis-regulatory elements important for the expression, but due to the specific site and number of integrations the expression intensity can vary (Gong et al., 2002).

Mouse lines with a knock-in of a recombinase at a specific genetic locus can be used as tools for cell lineage tracing as well as for the creation of cell population specific knock-outs that can result in a better understanding of signalling processes (see figure 12; Anderson et al., 2008). Crossing Cre lines to reporter lines that conditionally express fluorescent markers or β-galactosidase allow for the lineage tracing of cells that at one stage express Cre because the cells expressing Cre as well as their entire progeny will express the reporter gene (see figure 12b). The investigation of pathways and networks involved in a specific cell population that is marked by the expression of Cre can be carried out using conditional mouse lines in which one important gene for a certain pathway is irreversibly deleted in cells that express Cre (see figure 12c).

With this approach two Cre lines were generated as knock-ins of the CRF of Cre into the genetic locus of *Sox17* and *Foxa2* in order to express Cre most accurately within *Sox17* and *Foxa2*-positive cell populations in the endoderm. These mouse lines were analyzed by crossing them to a *ROSA* reporter mouse line that expresses β-galactosidase from the *ROSA* locus whenever Cre recombination has occured. β-galactosidase is expressed under the control of an ubiquitously active promoter when the Stop-codon flanked by *loxP* sites is eliminated by recombination.

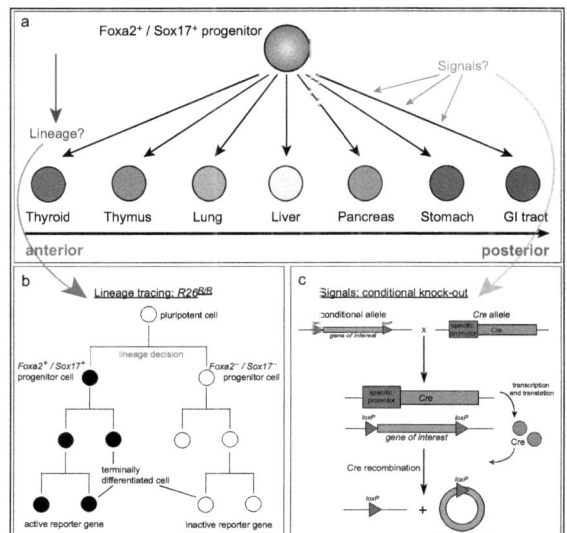

Figure 12:
Lineage tracing – identifying the progeny of progenitor cells – and studying the impact of signalling pathways – conditional knock-out analysis

The figure illustrates that there is a progenitor population of *Foxa2* and/or *Sox17* positive progenitor cells that gives rise to a certain portion of the endoderm-derived organs (thyroid, thymus, lung liver, pancreas stomach, gastro-intestinal tract) along the anterior-posterior axis while underlying different unknown differentiation signals (a). To investigate to what organs these cells will give rise to in the adult organism genetic cell lineage tracing is the most accurate approach that can be performed (a, b). Irreversible conditional activation of a reporter gene marks all cells expressing the recombinase under the transcriptional control of *Foxa2* or *Sox17* (Foxa2-Cre and Sox17-Cre lines) and their progeny (b). To functionally analyse the signals necessary for proper development of certain lineages that express either *Foxa2* or *Sox17*, genes can be conditionally deleted within these populations using the same Cre lines (carrying the Cre-allele = Cre is expressed under the transcriptional control of a certain gene – illustrated as a blue box; c). The same mechanism that activates reporter gene expression (by deletion of a Stop-codon placed right after the transcriptional start; b) can be use to eliminate one (heterozygous) or both alleles (knock-out) of a certain gene of interest (light blue) whose sequence is flanked by *loxP* sites (red arrowheads). Cre recombinase (orange octagons) recognizes *loxP* sites and cuts out sequences in between by leaving a single *loxP* site in the genome if the *loxP* sites have the same orientation (c).

(Abbreviations: GI tract = gastro-intestinal tract, *loxP* = locus of X over in P1)

Results

4.1. Cell lineage tracing - Generation and analysis of Cre mouse lines

4.1.1. Targeting of the *Sox17* locus

Sox17 is one of the first genes known to be expressed in the endoderm of the early mouse embryo (Kanai *et al.*, 1996). However it is not known to what cell types and organs *Sox17*-positive cells give rise to in the adult organism.

To address this question a Cre mouse line under the transcriptional control elements of *Sox17* was generated. Targeted mutagenesis in embryonic stem (ES) cells was used to introduce a codon improved Cre recombinase (improved Cre = iCre; Shimshek *et al.*, 2002) into exon 1 of the *Sox17* gene (see figure 13).

4.1.1. a) Design and generation of the targeting vector for the *Sox17* locus

The targeting vector was designed as shown in figure 13. It contains a mini gene locus, which consists of the *iCre* cDNA with a translational stop codon (see figure 13; red arrow), followed by a splice-donor site, an artificial intron and an exon encoding for the SV40 (simian virus 40) polyadenylation signal sequence (see figure 13; yellow and orange arrows, respectively). An *FRT*-flanked PGK (phospho-glycerate kinase) promoter-driven *neomycin* (*neo*) resistance gene was then placed after the polyadenylation signal in the opposite orientation (see figure 13; white arrow plus grey boxes marking *FRT* sites). The detailed generation of the targeting vector is described in the "Material and Methods" section (see Material and Methods, 6.2.14.).

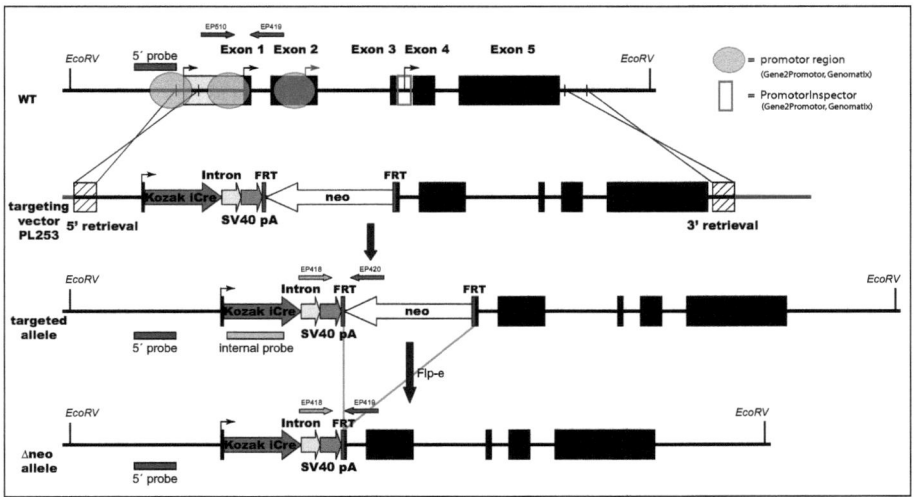

Figure 13: Targeting strategy of the *Sox17*^{iCre} allele

A targeting vector was used to introduce iCre into exon 1 of the *Sox17* gene and the selection cassette was removed by Flp-mediated excision. *Sox17* exons (black boxes) are numbered. Primers used for genotyping are designated EP418, EP419, EP421 and EP510. The external 5' probe and the internal Cre probe for Southern analysis are as indicated on the figure. Restriction enzyme sites for *EcoRV* are shown. Homology regions to generate the targeting constructs are indicated as 5' and 3' retrieval.
(Abbreviations: *FRT* = sites of Flp-recombination; iCre = improved Cre; neo = neomycin resistance gene; SV40pA = Simian Virus 40 polyadenylation signal)

4.1.1. b) Targeting of ES cells for homologous recombination and deletion of the selection cassette for the *Sox17* locus

TBV-2 ES cells (Wiles *et al.*, 2000) were electroporated with an *AscI*-linearized pL254 *Sox17*-iCre targeting vector and neomycin resistant clones were selected using 300µg/ml G418 (Invitrogen, 50mg/ml). Homologous recombination at the *Sox17* locus was confirmed by Southern blot analysis of *EcoRV*-digested genomic DNA using the *Sox17* 5'-probe (744bp; see figure 13 and Material and Methods, 6.2.13.) located outside of the targeting vector testing for locus-specific integration of the targeting vector (*Sox17* 5'-probe forward primer, including *HindIII* site; *Sox17* 5'-probe reverse primer including *AgeI* site). An internal probe was used to test for random or multiple integrations (restriction digest of the *iCre* sequence subcloned in pBSK- using *XmaI* and *NcoI*; for sequence etc. see Material and Methods). 2 out of 453 homologously recombined ES cell clones could be confirmed and were injected into recipient blastocysts (C57Bl/6) and chimeras and germline transmission of the *Sox17iCre* allele were obtained for both clones.

The *Sox17iCre/+* mice were mated to Flp-e deleter mice (Dymecki, 1996) on C57Bl/6 background to excise the neomycin selection cassette flanked by *FRT* sites using Flip recombination activity in the germline (for Southern blot verification see figure 14). The offspring were genotyped for both, the *Cre* and the *Flp-e* allele (see Material and Methods), and mice heterozygous for both alleles were backcrossed to C57Bl/6. Only mice that were negative for the *Flp-e* allele but that carried the *Cre* knock-in with the neomycin deletion were used for further analysis.

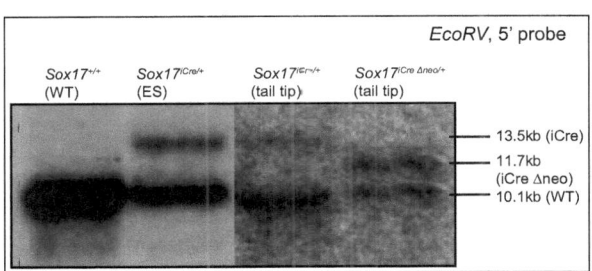

Figure 14:
Southern Blot on *Sox17* targeted ES cells and mice - Verification of homologously recombined clones
Southern blot of ES cell (lane 1 and 2) and mouse tail DNA (lane 3 and 4) digested with *EcoRV* and hybridized with the external 5' probe, showing the wild type allele (10.1kb), recombined allele (13.5kb) and neomycin deleted allele (11.7kb).

4.1.2. Analysis of the *Sox17iCre/+* mouse line

4.1.3. Recombination activity of the Sox17-iCre recombinase at early embryonic stages and after birth

For determination of the spatial and temporal pattern of the recombination activity of the iCre of *Sox17iCre/+* mice the mice were crossed to *R26R/R* (Soriano, 1999), a reporter line conditionally expressing β-galactosidase from the ROSA locus, to obtain mice heterozygous for both alleles (see figure 12b). In *R26R/R* mice β-galactosidase is expressed under the control of an ubiquitously active promoter when the Stop-codon flanked by *loxP* sites is irreversibly eliminated by Cre recombination (see figure 12b, for the principle of Cre-mediated recombination also see figure 12c). In a mouse that carries both the *Cre* and the *ROSA* reporter alleles all cells in which Cre recombination occurred and all their daughter cells can therefore be stained for β-galactosidase (see figure 12b).

Embryos and newborn pups of the *R26R/R* reporter mice were analyzed by β-galactosidase staining for the Cre recombination activity (see Material and Methods).

In embryos at E9.5 hardly any staining was detected in the endoderm. A very spotty staining that cannot be localized to a specific organ is spread all over the embryo and no more staining could be found using even the most stringent staining protocol (37°C, ON; see figure 15). The single stained cells that result in a spotty staining cover the head region, the heart, branchial arches and the region of the dorsal aorta (see figure 15).

The results for analysis of β-galactosidase activity obtained from organs (namely lung, stomach and heart) of young born mice (P1 = postnatal day 1) were in line with those obtained from early embryos. There was a strong staining detectable in the coronary vessels of the heart (see figure 16a and 16a', green and red arrows), the pulmonary artery and the aorta (see figure 16a and 16a", green and red arrows), in the pulmonary vessels (see figure 16b and 16b', green and red arrows) in the vessels lining the stomach and the bladder (see figure 16c, 16c' and 16d, green and red arrows). There was hardly any staining found in the endoderm derived organs except a

Results

chimaeric expression of β-galactosidase in the stomach (see figure 16c, 16c' and 16c", green and red arrows). Paraffin sections of the heart, lung and stomach (see figure 16a, 16a', 16b', 16c', 16c" and 16c''') revealed that the endothelium lining the vessels is positive for staining of β-galactosidase (see figure 16a, 16a', 16b' and 16c', green arrows) and, in the case of the stomach, also the epithelium lining the stomach cavity (see figure 16c" and 16c''').

The detailed expression analysis was further carried out by Perry Liao as part of his PhD thesis ("Generation of a Mouse Line Expressing *Sox17*-driven Cre Recombinase with Specific Activity in Arteries", Liao et al., Genesis, 2009).

Summarizing the expression data obtained with the *Sox17*$^{iCre/+}$ mouse line, strong Cre expression can be detected in the endothelium of the arteries and is therefore a valuable tool for recombination in theses cells but not in the endodermally derived organs.

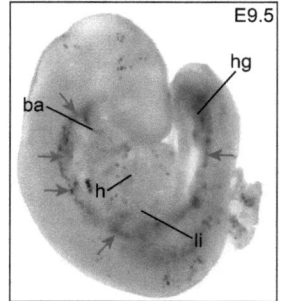

Figure 15:
Recombination activity of *Sox17*^{iCre} in the *ROSA* locus at E9.5
The β-galactosidase activity at E9.5 shows a very spotty scattered staining covering the whole embryo, especially in heart regions (h) and along the anterior-posterior axis in the dorsal aorta (red arrows).
(Abbreviations: ba = branchial arch; h = heart; hg = hindgut; li = liver)

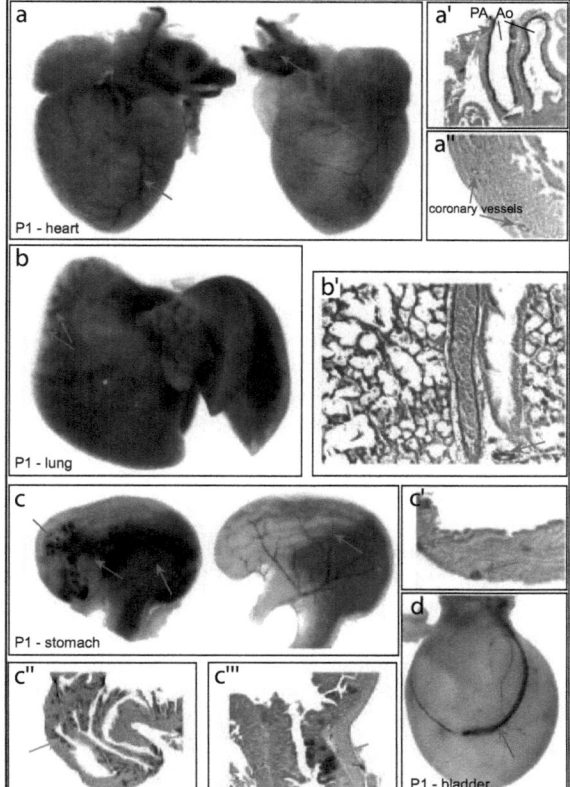

Figure 16:
β-galactosidase activity in organs of *Sox17*$^{iCre/+}$; *R26*$^{R/+}$ mice at P1

β-galactosidase activity in the heart (a; red arrows) can only be detected in the epithelium lining the arteries (a': pulmonary artery and aorta; a": coronary vessels; green arrows). In the lung recombination activity can also be detected in the vessels (b, b'; red and green arrows). In the stomach staining for β-galactosidase is apparent in the epithelium of the vessels (c, c'; red and green arrows) as well as in the epithelium lining the stomach cavity (yellow arrows; c", c''') and in the vessels along the bladder (d).
(Abbreviations: Ao = dorsal aorta; PA = pulmonary aorta; P1 = postnatal day 1)

4.1.4. Targeting of the Foxa2 locus

Foxa2 is a transcription factor specificially expressed in the early endoderm, the floorplate, the node and the notochord (Ang and Rossant, 1994; Weinstein et al., 1994). Its deletion causes severe abnormalities and failure in gastrulation, defects in neural tube patterning and gut morphogenesis that result embryonic lethality. Theses wide-spread effects are mainly due to the fact that Foxa2 expression is indispensible for the formation of organizer cell populations throughout gastrulation. Fate mapping studies revealed that the dynamic cell population of the gastrula organizer mainly contributes to prechordal mesoderm, midline and floorplate, cranial mesenchyme, the heart and the anterior endoderm as well as to somites, lateral mesoderm and notochord in a stage dependent manner (Kinder et al., 2001).

However it is still unknown to what organs and cell types Foxa2-positive cells of the early embryo give rise to in the adult mouse. To be able to exactly lineage trace these cells and to have a valuable tool to conditionally delete genes in the expression domain of Foxa2 (see figure 12c) a second Cre mouse line under the transcriptional control elements of the Foxa2 gene was generated. Targeted mutagenesis in embryonic stem (ES) cells was again used to introduce a codon improved Cre recombinase (iCre; Shimshek et al., 2002) into exon 1 of the Foxa2 gene, analogous to the $Sox17^{iCre/+}$ mouse line (see figure 17, Uetzmann et al., 2003).

4.1.4. a) Design and generation of the targeting vector for the Foxa2 locus

The targeting vector used for recombination in ES cells was designed as demonstrated in figure 17, according to the knock-in construct for Sox17.

It also contains a mini gene locus, consisting of the iCre cDNA with a translational stop codon (see figure 17; red arrow), followed by a splice-donor site, an artificial intron and an exon encoding for the SV40 polyadenylation signal sequence (see figure 17; yellow and orange arrows, respectively). The FRT-flanked PGK promoter-driven neo resistance gene follows the polyadenylation signal in the opposite orientation (see figure 17; white arrow plus grey boxes marking FRT sites). The detailed generation of the targeting vector for Foxa2 is described in the "Material and Methods" section (see Material and Methods, 6.2.15.).

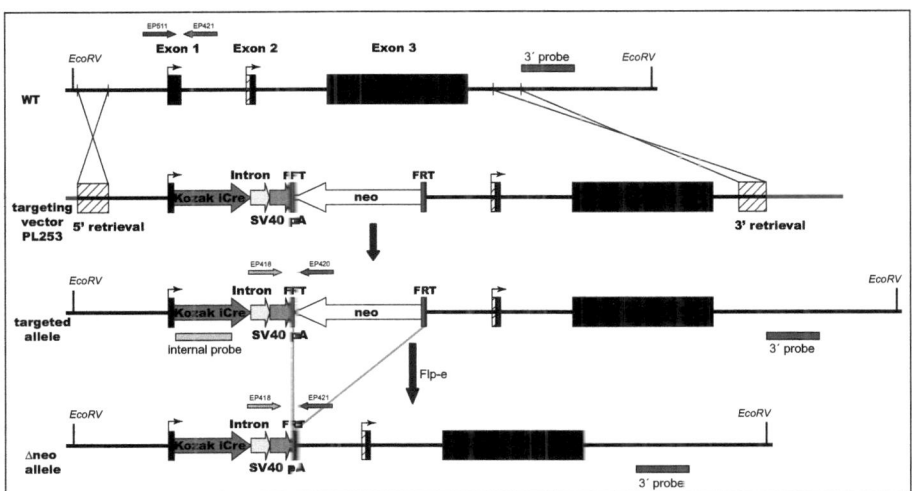

Figure 17: Targeting strategy of the $Foxa2^{iCre}$ allele

A targeting vector was used to introduce Cre into exon 1 of the Foxa2 gene and the selection cassette was removed by Flp-mediated excisions. Foxa2 exons (black boxes) are numbered. The alternative exon 2 (hatched box) and the two transcriptional starts (arrows) are indicated. Primers used for genotyping are designated EP418, EP420, EP421 and EP511. The external 3' probe and the internal probe are indicated. Restriction enzyme sites for EcoRV are shown. Homology regions to generate the targeting constructs are indicated as 5' and 3' retrieval.

4.1.4. b) Targeting of ES cells for homologous recombination for the *Foxa2* locus

To accomplish homologous recombination in mouse ES cells the *NotI*-linearized pL253 *Foxa2*-iCre targeting vector was transferred into TBV2 ES cells (Wiles et al., 2000) via electroporation. Neomycin resistant clones were selected using 300µg/ml G418 (Invitrogen, 50mg/ml). Homologous recombination specifically at the *Foxa2* locus was confirmed by Southern blot analysis of *EcoRV*-digested genomic DNA using the *Foxa2* 3´-probe (744bp; generated by PCR; for sequence see Material and Methods) located outside of the targeting vector and the same internal probe used for the *Sox17^{iCre}* knock-in to exclude random and multiple integrations (for generation see above; for sequence see Material and Methods). This way 2 out of 153 neomycin (neo) resistant clones were isolated as homologous recombinants and confirmed (see figure 18, lane 1 and 2). Two independent ES cell clones were injected into recipient blastocysts (C57Bl/6) to obtain chimeras and germline transmission of the *Foxa2^{iCre}* allele. The *FRT*-flanked *neo* selection cassette was deleted in the germline by Flip recombinase-mediated excision using Flp-e mice on a C57Bl/6 background (see figure 18, lane 3 and 4; Dymecki, 1996). After deletion mice were backcrossed to C57Bl/6 to eliminate the *Flp-e* allele. Only those mice negative for the *Flp-e* allele were used for backcrossings to C57Bl/6 and further analyses were carried out with *Foxa2$^{iCreΔneo/+}$* mice backcrossed to C57Bl/6 for five generations.

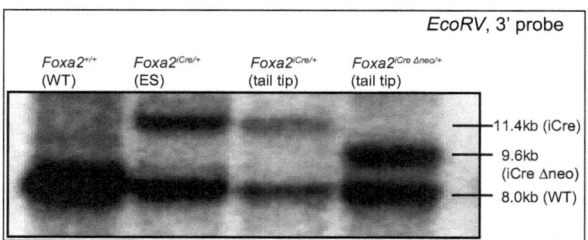

Figure 18:
Southern Blot on *Foxa2* targeted ES cells and mice - Verification of homologously recombined clones
Southern blot of ES cell (lane 1 and 2) and mouse tail DNA (lane 3 and 4) digested with *EcoRV* and hybridized with the external 3' probe, showing wild type allele (8.0kb), recombined allele (11.4kb) and neomycin deleted allele (9.6kb).

4.1.5. Analysis of the *Foxa2$^{iCre/+}$* mouse line

4.1.6. Recombination activity of the Foxa2-iCre recombinase at early embryonic stages

To determine the spatial and temporal patterns of Foxa2-iCre-mediated recombination of the *ROSA26* reporter (*R26R*) locus, we crossed our iCre mouse line to *R26R* mice (Soriano, 1999) to generate embryos heterozygous for both alleles. The expression of Cre under the transcriptional control of *Foxa2* results in recombination of the *ROSA* reporter allele leading to β-galactosidase expression driven by a ubiquitiously active promoter. The expression of β-galactosidase is then detected by its enzymatic reaction (using the LacZ-staining method).

This way recombination of the *R26R* allele in the visceral endoderm between E6.5 and E7.0 could not be observed (data not shown); as compared to the endogenous mRNA expression (Monaghan et al., 1993; Sasaki and Hogan, 1993; Ang and Rossant, 1994).

At E7.5 β-galactosidase activity was detected in cells anterior to the node and in the anterior endoderm (see figure 19a). At E8.5 (7-8 somites) Foxa2-iCre-mediated recombination was clearly detected in cells of the foregut and hindgut pocket as well as strongly in cells of the regressing node (see figure 19b). At E9.5, the β-galactosidase activity faithfully reflected the mRNA expression pattern of *Foxa2* in the floorplate, notochord, fore-, mid- and hindgut endoderm, as well as in the liver primordium (see figure 19c). Cre-mediated recombination of the *R26R* could also be shown in the first branchial pouch as well as in the developing pharynx.

At E10.5, β-galactosidase activity was detected in the anterior foregut, in the region of the forming lung and liver primordium, and continuously along the anteroposterior axis in the floorplate of the neural tube and in the notochord (see figure 19d). Foxa2-iCre-mediated recombination of the *R26R* was also detected in the first branchial pouch giving rise to the auditory tube. Histological sections confirmed Cre activity in the gut, liver and lung epithelium, as well as in the floorplate and notochord at different positions along the anteroposterior axis (see figure 19e and 19f).

Taken together, β-galactosidase activity in embryonic tissues faithfully reflected the expression of endogenous mRNA (Monaghan et al., 1993; Sasaki and Hogan, 1993; Ang and Rossant, 1994).

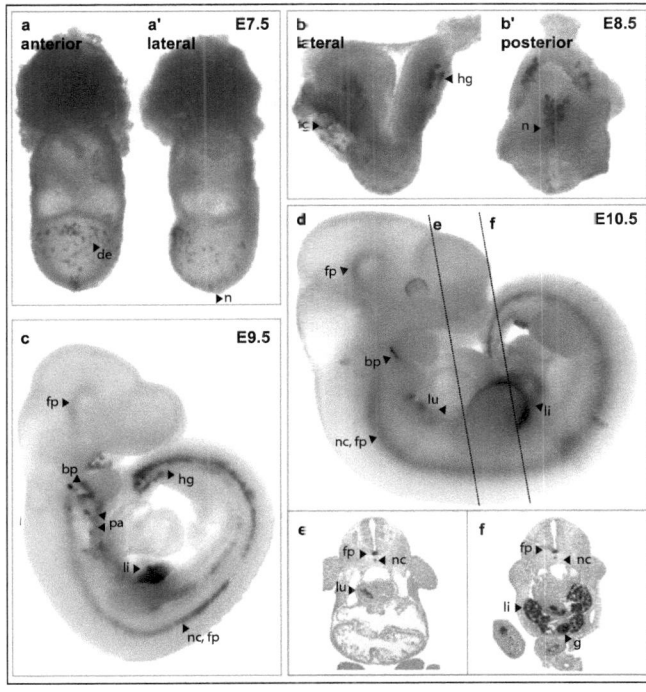

Figure 19:
Foxa2-iCre recombination activity at embryonic stages

Embryonic expression of *Foxa2*^{iCreΔneo} allele monitored by the *R26* reporter activity. (a) At E7.5 (80x), β-galactosidase activity is detectable in cells anterior to the node (n) and in the anterior definitive endoderm (de). At E8.5 (80x), iCre recombination is confined to the anterior foregut (fg) pocket, hindgut (hg) pocket, and the regressing node (n). (c) At E9.5 (20,8x), β-galactosidase activity is detected in the floorplate (fp), notochord (nc) and gut tube along the anteroposterior axis, as well as in the liver primordium (li), the branchial pouch (bp) and in the pharyngeal arches (pa). (d) At E10.5 (13x), β-galactosidase activity is detected in the same regions as at stage E9.5. e and f) Histological sections level indicated in (d) showing β-galactosidase activity in the floorplate, notochord, and lung (lu) primordium (e) and in the floorplate, notochord, gut epithelium (g) and liver (f).

4.1.7. Recombination activity of the Foxa2-iCre recombinase in embryonic organs

Between E9.5 and E12.5 different organs form along the anteroposterior axis of the primitive gut tube and to figure out to what extent *Foxa2* positive precursor cells contribute to these organs and whether it was possible to detect Foxa2-iCre-mediated recombination in these endoderm-derived lineages, organs at different stages of development were analysed.

To detect Cre-reporter activity in these endoderm-derived organs, organs from *Foxa2*^{iCre/+}; *R26*^{R/+} E12.5, E14.5, and E16.5 embryos were dissected and stained for β-galactosidase activity.

At E12.5, the liver, pancreas and GI tract show uniform β-galactosidase activity in the endodermal epithelium of these organs, while the expression of β-galactosidase in the epithelium of the lung appears to be slightly mosaic (see figure 20a). At E14.5 and E16.5, liver, stomach, and pancreas continue to show high β-galactosidase activity, whereas the lung shows clear mosaic expression of the β-galactosidase reporter (see figure 20a-d). Interestingly, the heart also shows reporter activity at E16.5 in over 50% (4 out of 7) of all animals (see figure 21a-c).

Histological sections were generated to investigate the β-galactosidase activity on a cellular level in the endoderm-derived organs between E12.5-16.5
The analysis revealed that the Foxa2-iCre recombination activity can be detected in all precursors and differentiated cells of the liver, pancreas and the gut epithelium and almost all cells of the epithelium lining the stomach cavity (see figure 22). Picture 22c clearly demonstrates that the pancreatic mesenchyme of the E14.5 embryo is negative for β-galactosidase staining and that the expression is restricted to endoderm-derived cells, progenitors of the exocrine and endocrine pancreas.

Results

It is also interesting to note that the β-galactosidase seems strongly activated in the crypt stem cell compartment as well as in all differentiated cell types of the gut epithelium at E16.5 (see figure 22d'). In contrast, the lung epithelium from E12.5 onwards shows mosaic staining for β-galactosidase (see figure 22e and 22e').

Histological sections of the heart revealed that the cardiac epithelia of both ventricles and atria are positive for β-galactosidase activity and confirmed the observed whole mount staining (see figure 21a').

In summary, theses results strongly indicate that Foxa2-iCre marks an early progenitor cell population of the endoderm that gives rise to a majority of the endoderm-derived organs and partially to the cardiac mesoderm, in which *Foxa2* mRNA expression cannot be detected.

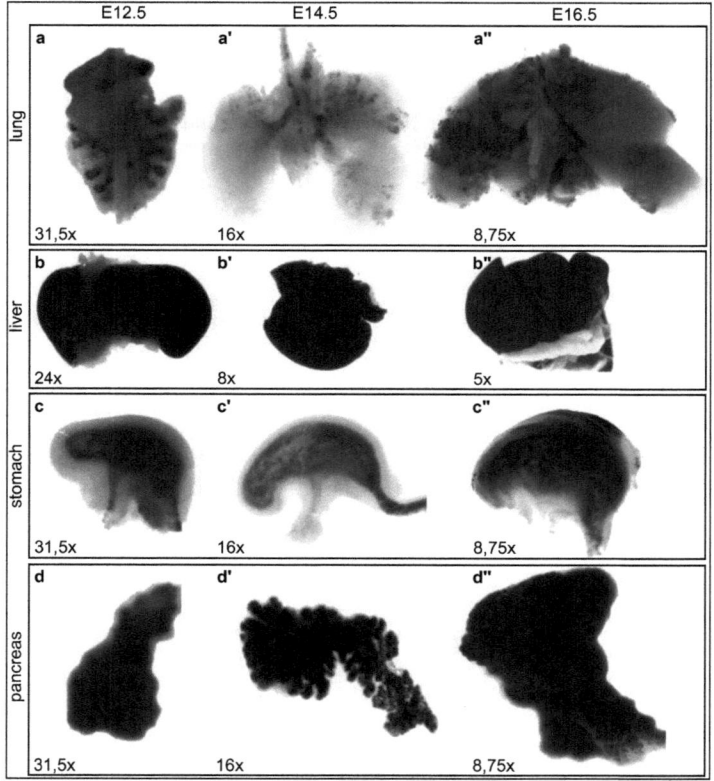

Figure 20: Recombination activity of Foxa2-iCre in embryonic organs

Cre recombination activity of *Foxa2*$^{iCre\Delta neo/+}$ mice monitored by the *R26R* in whole-mount organs. (a) Liver, (b) lung, and (c) stomach show high β-galactosidase activity from E12.5 to E16.5. (d) The lung epithelium shows a mosaic β-galactosidase activity in the epithelium from E12.5 to E16.5.

Figure 21: Recombination activity of Foxa2-iCre in the embryonic heart

Cre recombination activity in heart of *Foxa2*$^{iCre\Delta neo/+}$ mice reported by the expression of β-galactosidase from the *ROSA* locus. (a) E16.5 heart, lung and thymus show chimaeric activity of β-galactosidase. (a') Section through a heart of an E16.5 embryo. The epithelium lining the ventricle(s) and the atria shows Cre recombination activity monitored by the *R26R*. (b) E14.5 heart with starting β-galactosidase activity, (c) E15.5 heart with clearly positive atria for β-galactosidase activity.

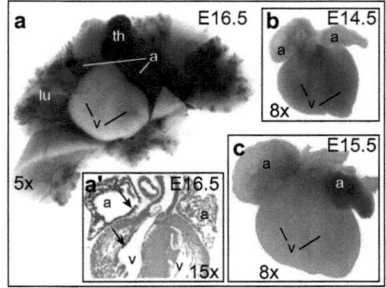

Results

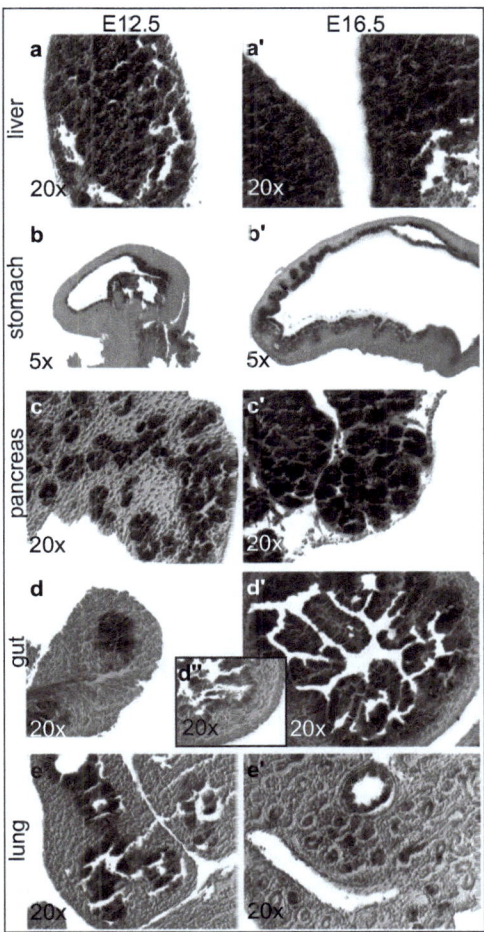

Figure 22:
Histological sections for β-galactosidase activity of organs of *Foxa2*[iCre/+]; *R26*[R/+] mice

The liver shows strong β-galactosidase activity at E12.5 (a) and E16.5 (a'). In the stomach Cre recombination activity is restricted to the stomach epithelium at E12.5 (b) and E16.5 (b'). The pancreatic epithelium shows uniform β-galactosidase activity at E12.5 (c) and E16.5 (c'), but not in the mesenchyme. In the gut epithelium all cells are positive for β-galactosidase activity at E12.5 (d) and E16.5 (d'). Endogenous β-galactosidase activity at E16.5 is shown in a wild type control (d''). In the lung epithelium at E12.5 (e) most cells are β-galactosidase positive, but show clear mosaic β-galactosidase expression at E16.5 (e').

4.2. Characterization of the *Foxa2*[iCre] hypomorphic allele

Foxa2[iCre/+] mice were mated to either *Foxa2*[iCre/+] or to C57Bl/6 to verify the anticipated null mutation (embryonic lethality around E10-11) as reported for *Foxa2* knock-out mice (Ang and Rossant, 1994; Weinstein et al., 1994) and the predicted heterozygous phenotype (jaw malloclusions etc.; Ang and Rossant, 1994). Neither of the expected results could be found regardless of how many generations of backcrossings to C57Bl/6 background were made. In addition to that homozygous mice were born and had no striking phenotype (in respect to behaviour, appearance, size and arrangement of organs etc.). 72 animals were genotyped 3 weeks after birth. Only 1 carried two *Foxa2*[iCre] alleles, 19 had only wild type alleles inherited and 42 were heterozygous (see chart 1 for results of Mendelian distribution). The Fisher's test (a X^2-square-test for independence vs. negative association with low numbers) for [11, 21, 21, 19] results in a left p value of $p \approx 0.097$ (Fisher, 1922 and 1954). The left p value is used as a confirmation of the probability of the parameters to have negative influence on each other. From this it follows that the distribution does not follow a Mendelian distribution, but is lower, with an error probability of ~10%. This implies that mice homozygous for the iCre allele die before P30, either due to embryonic or early postnatal lethality.

To analyse the level of Foxa2 protein and determine its concentration in *Foxa2* wild type versus *Foxa2*[iCre/+] and

Results

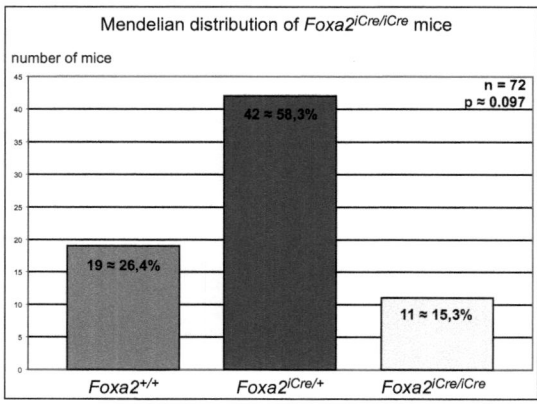

Chart 1:
Results of Mendelian distribution of $Foxa2^{iCre/iCre}$ mice
The chart shows numbers for $Foxa2^{+/+}$, $Foxa2^{iCre/+}$ and $Foxa2^{iCre/iCre}$ mice from 11 different litters. The total numbers of mice of all litter is indicated (n = 72) as well as the corresponding p-value for the distribution (p ≈ 0.097); total numbers of each genotype are specified each column (lilac = $Foxa2^{+/+}$; purple = $Foxa2^{iCre/+}$; light yellow = $Foxa2^{iCre/iCre}$). Calculated percentages are indicated within each column.

$Foxa2^{iCre/iCre}$ western blot analysis on P1 liver was performed. Protein lysates from P1 homozygous mutant showed that 10-15% of the Foxa2 protein remained in comparison to wild type. The $Foxa2$ gene expression in $Foxa2^{iCre/+}$ liver at P1 is also reduced to 40-50% compared to the wild type sample (see figure 23).
As revealed by Western blot analysis, $Foxa2^{iCre/iCre}$ mice still express $Foxa2$ mRNA which is translated into the

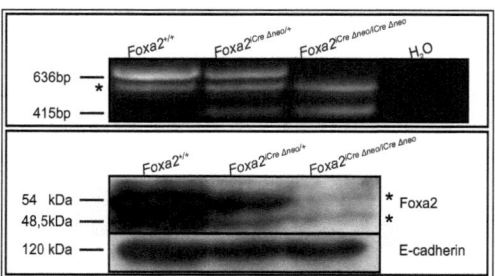

Figure 23:
Analysis of Foxa2 protein levels in wild type versus $Foxa2^{iCre/+}$ and $Foxa2^{iCre/iCre}$ liver extracts at P1 using Western blot
(a) PCR genotyping of $Foxa2^{iCre/+}$ mice using primers EP511, EP421 and EP418 to distinguish between the wild type allele (636bp) and $Foxa2^{iCreΔneo}$ allele (415bp). Asterisk indicates a non-specific PCR product of approximately 500bp. (b) Western blot analysis of P1 mouse liver lysates. Foxa2 protein is reduced in $Foxa2^{iCreΔneo/+}$ samples and almost not detectable in $Foxa2^{iCreΔneo/iCreΔneo}$ liver lysates. The two protein bands indicated by asterisks most likely correspond to Foxa2 unmodified and post-translationally modified Foxa2 protein. E-cadherin was used as a loading control.

full length protein, although at reduced levels. To understand why the targeting strategy failed to generate a null mutation the regulation of the $Foxa2$ gene expression was further investigated using the predictive Genomatix software package (Eldorado Release 4.5, Genomatix Software GmbH).
The software predicted a second promoter downstream of the known $Foxa2$ promoter located in intron 1, supported by 30 CAGE (Cap Analysis Gene Expression; Shiraki et al., 2003) tags (see figure 24). From the alternative transcriptional start site upstream of exon 2 a mRNA encoding the complete open reading frame of $Foxa2$ is transcribed and is also supported by the prediction of the Human And Vertebrate Analysis and Annotation (HAVANA) group of the Sanger Institute (see figure 24 and ENSEMBL). In summary, Foxa2 protein expression is based on the translation of two transcripts that only differ in the 5' UTR driven by two promoters.

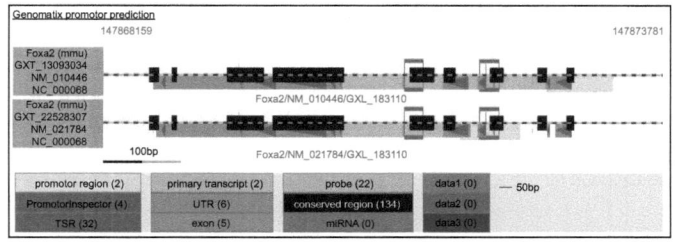

Figure 24:
Genomatix promoter prediction for $Foxa2$

Promoter regions (yellow boxes) are predicted for two different transcripts as also indicated in figure 17 (arrows for transcriptional start sites). Exons and UTR are shown in green, transcriptional start regions (TSR) are designated as red arrows in pink boxes. Conserved

region between different species are depicted in purple boxes. Grey boxes show the primary transcript.
(Abbreviations: TSR: transcriptional start region; UTR: untranslated region)

Since the mice homozygous for the *iCre* allele are viable and since they show a chimaeric expression of β-galactosidase in several tissues (e.g. the lung; see figure 20a) when crossed to the *R26* reporter line, lungs and livers of adult mice were analysed by RT-PCR and Western blot, in regard to their expression of the *iCre* and *Foxa2* mRNA and protein (see figure 26 and 27) as a reflection of the activity of the two promoters identified.
The RT-PCR was performed on cDNA of DNase-treated total mRNA samples from lung and liver (see figure 26). The primers were designed as shown in figure 25 to qualitatively proof the transcription of all possible transcripts in the different tissues.

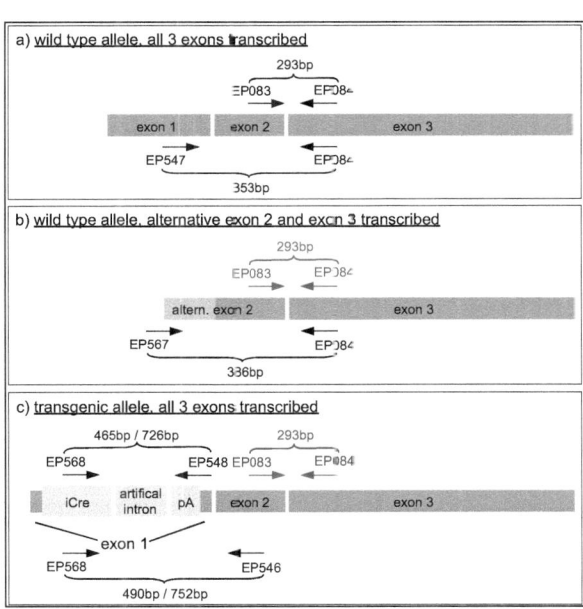

Figure 25:
Primers for RT-PCR on RNA samples from different tissues of *Foxa2*$^{iCre\Delta neo/iCre\Delta neo}$, *Foxa2*$^{iCre/+}$ and *Foxa2*$^{+/+}$ mice

Different primer combinations will prove the existence of different transcripts. The primers are designed as indicated in the figure. Localization of the primers are pointed out by arrowheads. Possible transcripts of the wild type allele are amplified using EP547 and EP084 (transcript of all three exons driven by promoter 1, 353bp), EP567 and EP084 (transcript driven by promoter 2 consisting of 5' elongated exon 2 and exon 3, 386bp) and EP083 and EP084 (amplification of both possible transcripts, 293bp). Possible transcript of the *iCre* allele are amplified using EP568 and EP548 (combination of primers amplifies the *iCre* allele only: 726bp with artifical intron – no correct splicing; 465bp without the artificial intron – correct splicing), EP568 and EP546 (for the amplification of fusions of the *iCre* to exon 2: 752bp with artifical intron; 490bp without the artificial intron+/- part 2 of exon 1).

The quality and quantity of the RNA and cDNA were tested by PCR on β-actin (amplified with EP142 and EP143; see figure 26, lane 6) that also served as a semi-quantitative loading control. The controls show that the ORF of *Foxa2* is transcribed in every sample (amplified with EP083 and EP084; see figure 26, lane 5). The exon 1 transcript was not detectable in the lung or in the homozygous sample of the liver; however it was detected in liver samples of heterozygous and wild-type mice (EP547 and EP084; see figure 26, lane 1).
The *iCre* transcript (amplified with EP568 and EP548; see figure 26, lane 2) could only be seen in the liver but not in the lung. The artificial intron was spliced out correctly because only the smaller amplified fragment could be detected. In line with this result, the alternative transcript driven by the second promoter downstream could be found in the lung for all genotypes. In liver, wild-type and heterozygous samples also show expression of that alternative transcript, but it could not be detected in the homozygote (amplified with EP567 and EP084; see figure 26, lane 3). A fusion of the *iCre* transcript to exon 2 could not be detected in any of the samples (amplified with EP568 and EP546; see figure 26, lane 4).

Results

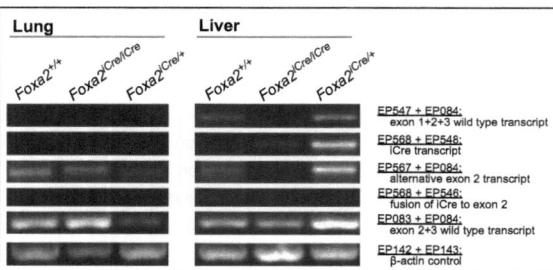

Figure 26:
Analysis of RNA samples from lung and liver of $Foxa2^{iCre\Delta neo/iCre\Delta neo}$, $Foxa2^{iCre\Delta neo/+}$ and $Foxa2^{+/+}$ mice regarding different possible transcripts using RT-PCR
RT-PCR on lung (left) and liver (right) samples of $Foxa2^{+/+}$ (first and fourth column) $Foxa2^{iCre\Delta neo/+}$ (second and fifth column) and $Foxa2^{iCre\Delta neo/iCre\Delta neo}$ (third and sixth column) was performed. β-actin was used as control for the integrity of the RNA and cDNA and as a semi-quantitative loading control (last row). Primer combination EP547 and EP084 amplifies the wild type exon 1 transcript (row 1; also see figure XX), EP568 and EP548 amplify the
iCre transcript (row 2; also see figure XX), EP567 and EP084 the alternative exon 2 transcript (row 3; also see figure XX), EP568 and EP546 the fusion of the iCre transcript to exon 2 and EP083 and EP084 the transcript of the ORF of Foxa2 (see figure XX).
(Abbreviations: ORF = open reading frame)

To analyse the Foxa2 protein levels in brain, lung and pancreas in wild type versus $Foxa2^{iCre/+}$ and $Foxa2^{iCre/iCre}$ adult (≥8 weeks) mice Western blot was performed. The results show a clear difference in the expression of Foxa2 in the pancreas, with less expression in animals which are heterozygous and homozygous for the iCre allele (see figure 27b). There is also a severe decrease of expression in the lung in wild type compared to heterozygous and homozygous animals (see figure 27a). Quantification using the gel analyzer tool of the "Image J" software revealed that the expression of Foxa2 in the lung in animals that are homozygous for the iCre allele is about 2/3 of the amount of Foxa2 protein concentration in wild type animals (1.23 → 0.85; see figure 27a). In the pancreas the decrease of Foxa2 protein in homozygous animals is about 1/2 compared to the wild type level (1.50 → 0.73; see figure 27b). Regarding the brain, protein levels were too low to detect any expression. The expression of Foxa2 in the adult brain seems to be restricted to a relatively small cell population in the ventral midbrain (Kittappa et al., 2007).
Determinations of the levels of mRNA expression and protein synthesis strongly indicate that the two promoters predicted by the Genomatix software are regulated in a tissue specific manner.

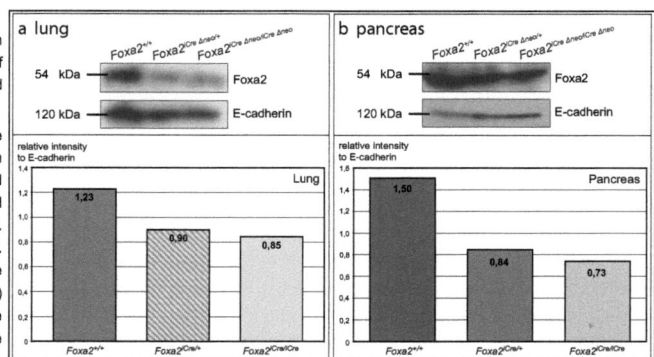

Figure 27:
Analysis of the Foxa2 protein concentration in samples from lung and pancreas of $Foxa2^{iCre\Delta neo/iCre\Delta neo}$, $Foxa2^{iCre\Delta neo/+}$ and $Foxa2^{+/+}$ mice using Western blot

In the lung (a) like in the liver of P1 mice (see figure 23) a decrease in Foxa2 protein expression can be detected between wild type (first lane) and heterozygous (second lane) and homozygous (third lane). E-cadherin serves as loading control. The decrease of Foxa2 expression in the pancreas (b) is as severe as in lung (a) and liver (see figure 23). The first lane shows expression of Foxa2 in the wild type pancreas, in the second and third lane
heterozygous and homozygous samples are shown, respectively. Again, E-cadherin is used as a loading control. The corresponding charts in a and b show the quantification of the Foxa2 protein expression levels normalized to E-Cadherin for $Foxa2^{+/+}$, $Foxa2^{Cre/+}$ and $Foxa2^{iCre/iCre}$ using the "Image J" gel analysis method. Exact relative values are indicated in the columns.
(Abbreviations: kDa = kilo Dalton; P = postnatal day)

4.3. Analysis of the metabolism of $Foxa2^{iCre\Delta neo/iCre\Delta neo}$ mice

The Foxa family of transcription factors has been shown to be involved in early embronic development (Ang and Rossant, 1994; Weinstein et al., 1994) and only recently its function in liver metabolism has begun to be described (Wolfrum et al., 2004; Wolfrum and Stoffel, 2006; Wolfrum et al., 2008). Foxa2 in particular is involved in insulin responsive liver gene expression, fatty acid metabolism and bile duct homeostasis (Wolfrum et al., 2004; Wolfrum and Stoffel, 2006; Wolfrum et al., 2008; Wederell et al., 2008, Bochks et al., 2008). Even *Foxa2* heterozygous mice show alterations in liver metabolite concentrations, e.g. high density lipoprotein (HDL; Wolfrum et al., 2008), in line with the observation that *Foxa2* heterozygotes show increased adiposity on a high-fat diet (Wolfrum et al., 2003).

The high levels of recombination activity of the iCre observed in *Foxa2*$^{iCre\Delta neo/+}$ mice in liver and pancreas led to the consideration of a possible metabolic defect in mice homozygous for the *iCre* allele, the effects of which would eventually appear in adulthood only.

Therefore littermates of heterozygous intercrosses (*Foxa2*$^{iCre\Delta neo/iCre\Delta neo}$; *Foxa2*$^{iCre\Delta neo/+}$ and *Foxa2*$^{+/+}$; generation 5 backcrossed to C57Bl/6 background) were analysed for evidence of metabolic issues approximately 8 weeks after birth in cooperation with Dr. Susanne Neschen (Helmholtz Zentrum München, Institute for Experimental Genetics). Mice were anaesthetized with isoflurane and blood from the *vena cava* was taken for plasma and blood analysis. Markers for liver metabolism were examined in a plasma analysis (plasma lipids, glucose etc.). For comparison of the plasma data, liver and body weight (BW) were measured and the liver weight was calculated in relation to the body weight (% BW). There was no significant variability in the liver size or weight compared to the body weight (see table 2, column 4).

The analysis of the blood showed a decrease in the hemoglobin concentration and the hematocrit in heterozygous and homozygous vs. wild type mice (see table 1, column 5 and 6), although the number of red blood cells was indiscernably lower.

Genotype	White blood cells	Red blood cells	Platelets	Hemoglobin concentration	Hematocrit
	1.000/µl	1.000.000/µl	1.000/µl	mg/dl	%
Foxa2$^{+/+}$	6,20	10,60	971	16,50	50,95
Foxa2$^{iCre/+}$	4,45	8,68	1003	13,16	41,13
Foxa2$^{iCre/iCre}$	6,12	9,03	1023	13,76	43,37

Table 1: Blood analysis of *Foxa2*$^{iCre\Delta neo/iCre\Delta neo}$, *Foxa2*$^{iCre\Delta neo/+}$ and *Foxa2*$^{+/+}$ mice
The table shows the results of the blood analysis of *Foxa2*^{iCre} mice. Row 1 refers to the wild type mouse row two and three refer to heterozygous and homozygous mice, respectively. Parameters measured are concentrations of white and red blood cells, of the platelets, of the hemoglobin and the hematocrit (column 2-6). Slight differences in the blood analysis from wild type vs. heterozygous and homozygous can already be detected. Heterozygous and homozygous samples show a decreased hemoglobin concentration and a lower hematocrit (column 5 and 6). The numbers of white and red blood cells as well as the platelet number do not differ significantly (column 2-4).

The plasma analysis revealed, that for all parameters measured *Foxa2*$^{iCre\Delta neo/+}$ and *Foxa2*$^{iCre\Delta neo/iCre\Delta neo}$ mice show a significant decrease compared to the wild type control mouse (cholesterol, triacylglycerol, glucose, HDL (high density lipoprotein), LDL (low density lipoprotein), non-esterified fatty acids; see table 2, column 5-10) while the percentage of the liver weight compared to the whole body weight, the body weight itself or the weight of the liver do not differ in wild type vs. heterozygous and homozygous mice (see table 2; column 2-4).

These results indicate that the *Foxa2* hypomorphic mice suffer from liver metabolic defects and suggest that these mice can be used to study signalling pathways involved in liver metabolism and associated defects.

Genotype	BW (g)	Liver weight (g)	Liver % BW	Cholesterol	Triacyl-glycerol	Glucose	LDL	HDL	Non-esterified fatty acids
Foxa2$^{+/+}$	21,2	1,2	5,7	132,8	204,2	107,2	22,4	90,4	2,3
Foxa2$^{iCre/+}$	21,7	1,2	5,5	94,8	124,6	100,6	16,2	65,6	1,3
Foxa2$^{iCre/iCre}$	20,6	1,2	5,8	84,4	129,8	81,4	18,3	55,4	1,4

Table 2: Plasma analysis of *Foxa2*$^{iCre\Delta neo/iCre\Delta neo}$, *Foxa2*$^{iCre\Delta neo/+}$ and *Foxa2*$^{+/+}$ mice
The table shows the results of the plasma analysis for *Foxa2*^{iCre} mice. Row 1 refers to the wild type mouse row two and three refer to heterozygous and homozygous mice, respectively. Parameters measured are cholesterol, triacylglycerol, glucose, LDL, HDL and non-esterified fatty acids (column 5-10). Heterozygous and homozygous mice show a decrease for all liver metabolic parameters in the analysis (cholesterol, triacylglycerol, glucose, HDL, LDL, non-esterified fatty acids; column 5-10 row 2 and 3). Column 2-4 show liver and body weight and the calculated liver weight in relation to the body weight in %. The body weight as well as the liver weight does not differ significantly between wild type, heterozygous and homozygous mice (column 2-4).
(Abbreviations: BW = body weight; HDL = high density lipoprotein; LDL = low density lipoprotein)

Results

4.4. Conditional knock-out analysis

Conditional deletion of genes that would usually lead to early embryonic lethality in the complete knock-out situation provides the chance to study the function of those genes at later stages or in specific tissues. Necessary tools for approaches like this are recombinases, either temporally restricted (inducible) and/or tissue-specific. The generated mouse lines, $Foxa2^{iCre}$ and $Sox17^{iCre}$, expressing Cre in a specific subset of cells could therefore be used as tools for conditional knock-out analysis.

To understand how endoderm progenitors become lineage-specified and how they give rise to different cell types and organs and, in particular, what signals are involved in these lineage decisions it is necessary to knock out key factors of the different pathways that should be tested for involvement (see figure 12a and 12c).

The impact of the canonical Wnt pathway, e. g., on any differentiation processes can be studied by elimination of its downstream effector β-catenin (see figure 8).

4.4.1. Characterization of the β-catenin knock-out in the Foxa2-positive cell population: $Foxa2^{iCre/+}$; β-catenin$^{flox/flox}$; $R26^{R/+}$ mice

The Nodal/TGFβ as well as the Wnt/β-catenin pathways are indispensible for the induction of mesoderm and endoderm; for both pathways knock-out models exist that fail to gastrulate and fail to form mesoderm and endoderm (Conlon et al., 1994; Zhou et al., 1993; Haegel et al., 1995; Liu et al., 1999; Huelsken et al., 2000). A gradient of Nodal signaling is sufficient to pattern mesoderm versus endoderm with high and low levels, respectively (Zhou et al., 1993; Conlon et al., 1994; Alexander and Stainier, 1999; Tremblay et al., 2000; Lowe et al., 2001; Vincent et al., 2003). The Wnt/β-catenin pathway plays a role in the separation of endoderm versus mesoderm (Lickert et al., 2002). β-catenin mutants, knock-out models for canonical Wnt signalling, in the domain of Cytokeratin 19 Cre (K19-Cre) expression show an accumulation of mesoderm in the expense of endoderm, or more precisely ectopic cardiac mesoderm is formed (Lickert et al., 2002).

To investigate the influence of β-catenin, in terms of canonical Wnt signalling, in the Foxa2-positive cell population in the early embryo, β-catenin was conditionally deleted under the control of the $Foxa2^{iCre}$ allele. Thus, deletion should occur in mesendodermal progenitor cells giving rise to mesoderm and endoderm. Therefore $Foxa2^{iCre/+}$ mice were mated to conditional β-catenin knock out mice, carrying the ROSA lacZ reporter gene (β-catenin$^{flox/flox}$; $R26^{R/R}$; Brault et al., 2001; Soriano, 1999), to obtain mice heterozygous for all three alleles (see figure 28; $Foxa2^{iCre/+}$; β-catenin$^{flox/+}$; $R26^{R/+}$). The reporter allele thereby allows for the sensitive detection of cells, cell populations and tissues with Cre recombination activity, most likely overlapping with the conditional deletion of β-catenin (see figure 12b and 12c).

As expected, deletion of just one allele of β-catenin resulted in viable mice that had neither an obvious phenotype nor any difference in β-galactosidase expression compared to wild type mice.

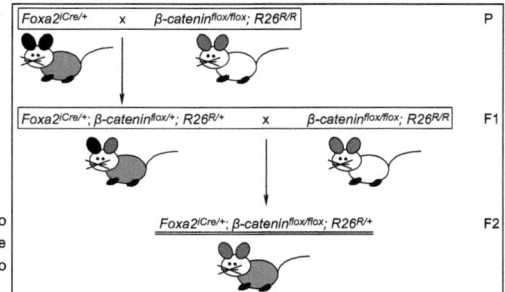

Figure 28: Crossing scheme conditional β-catenin knock out A
$Foxa2^{iCreΔneo/+}$ mice were crossed to β-catenin$^{flox/flox}$; $R26^{R/R}$ mice to obtain $Foxa2^{iCreΔneo/+}$; β-catenin$^{flox/+}$; $R26^{R/+}$ mice (F1). These mice were used for plug matings with β-catenin$^{flox/flox}$; $R26^{R/R}$ mice to analyze $Foxa2^{iCreΔneo/+}$; β-catenin$^{flox/flox}$; $R26^{R/+}$ mice (F2).

To completely knock out β-catenin in the $Foxa2^{iCre}$ expression domain $Foxa2^{iCre/+}$; β-catenin$^{flox/+}$ mice were crossed to β-catenin$^{flox/flox}$; $R26^{R/R}$ and the offspring was genotyped for mice carrying the iCre allele and two floxed β-catenin alleles (see figure 28).

No developmental defects or embryonic lethality could be observed between E6.5 and E18.5. Moreover, it was shown that the mice with two conditional alleles of β-catenin and the $Foxa2^{iCre}$ allele are viable and are born in an expected Mendelian ration of ~25% (28,5%, n=70, see chart 2).

The unexpected result that conditional deletion of β-catenin in the domain of Foxa2-iCre recombination activity does not lead to embryonic lethality suggests that the Cre recombination is not sufficient. At early stages recombination might occur too late, or the expression could be not strong enough for complete recombination, generating a genetic mosaic of wild type and conditional mutant cells.

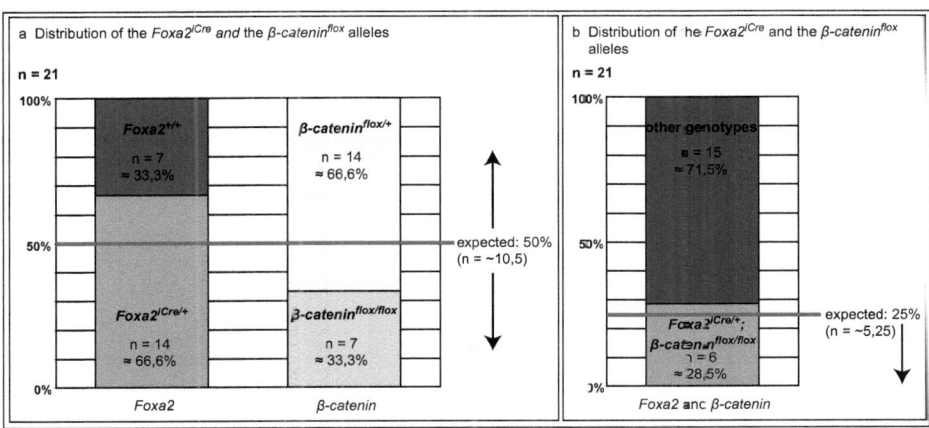

Chart 2: Preliminary data of Mendelian distribution of Foxa2[iCre/+]; β-catenin[flox/flox] mice

The chart shows the distribution of the different alleles within a mouse colony with 21 members. The Foxa2[iCreΔneo] allele is inherited with 66% the β-catenin[flox] allele with 33% (a). Mendelian distribution would be 50% for each (red line). However, together they are inherited with 28,5% (b), while the expected distribution is 25% (red line).

4.4.2. Analysis of the metabolism of Foxa2[iCreΔneo/+]; β-catenin[flox/flox]; R26[R/+] mice

Due to the fact that the mutation of β-catenin in Foxa2[iCreΔneo/+]; β-catenin[flox/flox]; R26[R/+] mice is not embryonic lethal the mice were analysed for a later metabolic phenotype. It was known from the data achieved by crossing Foxa2[iCreΔneo/+] to R26[R/R] β-galactosidase reporter mice that the Cre recombination activity is high in all cells of the liver and the pancreas. If canonical Wnt signaling has an influence on the metabolism of pancreas or liver it is most likely that the mutant mice exhibit a metabolic phenotype that can be revealed by measurement of blood plasma parameters.

Therefore eight week old litter mates (note: n=1 for each genotype) were anesthetized and blood was taken from the vena cava to obtain data from blood and plasma analysis. Again, to be able to compare the data, the body weight (BW) and the liver weight were measured. The liver weight was then calculated in relation to the body weight ("% body weight"). No differences could be found for the size or the weight of the liver (see table 4, column 4).

The results of the blood and plasma analysis revealed that there is no significant difference detectable regarding blood or plasma samples. Blood cell types are present in the expected distribution, number and concentration with higher variability for the Foxa2[iCreΔneo/+]; β-catenin[flox/flox] mouse (see table 3, column 2-6). Also liver metabolism specific parameters are within the normal range and do not differ significantly between "wild type" (Foxa2[+/+]; β-catenin[flox/+]), heterozygotes (Foxa2[iCreΔneo/+]; β-catenin[flox/+]) and homozygotes (Foxa2[iCreΔneo/+]; β-catenin[flox/flox]; cholesterol, triacylglycerol, glucose, HDL (high density lipoprotein), LDL (low density lipoprotein), non-esterified fatty acids; see table 4, column 5-10). Only two HDL and non-esterified fatty acids showed a lower concentration in the Foxa2[iCreΔneo/+]; β-catenin[flox/flox] mouse.

Results

Genotype	White blood cells	Red blood cells	Platelets	Hemoglobin concentration	Hematocrit
	1.000/µl	1.000.000/µl	1.000/µl	mg/dl	%
Foxa2$^{+/+}$; β-catenin$^{flox/+}$	2,71	9,33	1332	13,92	43,02
Foxa2$^{iCre/+}$; β-catenin$^{flox/+}$	1,81	9,11	1248	13,3	42,55
Foxa2$^{iCre/+}$; β-catenin$^{flox/flox}$	4,54	7,67	994	11,6	36,6

Table 3: Blood analysis of Foxa2$^{iCreΔneo/+}$; β-catenin$^{flox/flox}$ mice

The table shows the results of the blood analysis of Foxa2$^{iCreΔneo/+}$; β-catenin$^{flox/flox}$ mouse in comparison to Foxa2$^{iCreΔneo/+}$; β-catenin$^{flox/+}$ and Foxa2$^{+/+}$; β-catenin$^{flox/+}$ mice. Parameters measured are concentrations of white and red blood cells, of the platelets, of hemoglobin and the hematocrit (column 2-6). Only the Foxa2$^{iCreΔneo/+}$; β-catenin$^{flox/flox}$ mouse shows slight differences in the concentrations of all parameters measured (last row).

Genotype	BW (g)	Liver weight (g)	Liver (% BW)	Cholesterol	Triacyl-glycerol	Glucose	LDL	HDL	Non-esterified fatty acids
Foxa2$^{+/+}$; β-catenin$^{flox/+}$	21,8	1,0	4,6	112,2	128,0	116,2	19,1	77,0	1,9
Foxa2$^{iCre/+}$; β-catenin$^{flox/+}$	21,9	1,1	5,0	95,8	106,6	103,0	16,6	60,4	1,9
Foxa2$^{iCre/+}$; β-catenin$^{flox/flox}$	24,1	1,1	4,6	100,4	132,0	88,2	19,3	55,8	1,2

Table 4: Plasma analysis of Foxa2$^{iCreΔneo/+}$; β-catenin$^{flox/flox}$ mice

The table shows the results of the plasma analysis of Foxa2$^{iCreΔneo/+}$; β-catenin$^{flox/flox}$ mouse in comparison to Foxa2$^{iCreΔneo/+}$; β-catenin$^{flox/+}$ and Foxa2$^{+/+}$; β-catenin$^{flox/+}$ mice. Parameters measured are cholesterol, triacylglycerol, glucose, LDL, HDL and non-esterified fatty acids (column 5-10). Only the Foxa2$^{iCreΔneo/+}$; β-catenin$^{flox/flox}$ mouse shows slight differences in the concentrations of especially HDL and non-esterified fatty acids (last row). Column 2-4 show liver and body weight and the calculated liver weight in relation to the body weight in %. The body weight as well as the liver weight do not differ significantly between Foxa2$^{+/+}$; β-catenin$^{flox/+}$, Foxa2$^{iCreΔneo/+}$; β-catenin$^{flox/+}$ and Foxa2$^{iCreΔneo/+}$; β-catenin$^{flox/flox}$ mice (column 2-4).

4.4.3. Characterization of the ß-catenin knock-out in the Foxa2-positive cell population: Foxa2$^{iCre/+}$; β-catenin$^{floxdel/flox}$; R26$^{R/+}$ mice

Because mice carrying two floxed β-catenin and the Foxa2$^{iCre/+}$ alleles were viable and had no obvious phenotype, contrary to the predictions made on the basis of former published observations (Lickert et al., 2002), the number of alleles to be recombined was reduced, assuming that this way the deletion of β-catenin would be more complete and less of a genetic mosaic would be generated.

To achieve a reduction of the number of alleles for recombination, mice heterozygous for the β-catenin allele (β-catenin$^{floxdel/+}$; Brault et al., 2001) were mated to Foxa2$^{iCre/+}$ mice to obtain mice positive for both alleles (Foxa2$^{iCre/+}$; β-catenin$^{floxdel/+}$). These mice were than mated to β-catenin$^{flox/flox}$; R26$^{R/R}$ to completely delete β-catenin in the domain of Foxa2-iCre recombination activity by only one recombination step that in parallel could be visualized using the reporter activity of the R26$^{R/+}$ allele (see figure 12b, 12c and 29).

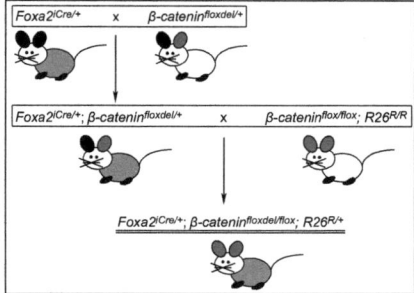

Figure 29: Crossing scheme conditional β-catenin knock out B

Foxa2$^{iCreΔneo/+}$ mice were crossed to β-catenin$^{floxdel/+}$ mice to obtain Foxa2$^{iCreΔneo/+}$; β-catenin$^{floxdel/+}$ mice (F1). These mice were used for plug matings with β-catenin$^{flox/flox}$; R26$^{R/R}$ mice to analyze Foxa2$^{iCreΔneo/+}$; β-catenin$^{floxdel/flox}$; R26$^{R/+}$ mice (F2).

Embryos dissected at E10.5 had no obvious phenotype. They had neither size abnormalities, regarding whole embryos or their organs, nor issues regarding the arrangement of the organs. There was no observable difference in their staining pattern for β-galactosidase activity as compared to control embryos (see figure 30). The mice were therefore mated to produce live offspring.

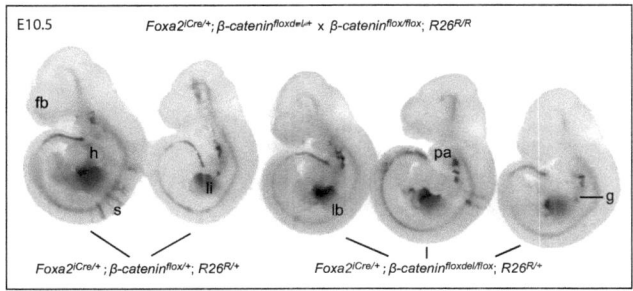

Figure 30:
E10.5 $Foxa2^{iCre/+}$; $\beta\text{-}catenin^{floxdel/flox}$; $R26^{R/+}$ mice

The figure shows 5 embryos dissected at E10.5. Yolk sacs were used for genotyping. Mice show activity of β-galactosidase in the expected tissues (e.g.: liver, pharyngeal arches, notochord and floorplate). Genotypes from left to right: $Foxa2^{iCre\Delta neo/+}$, $\beta\text{-}catenin^{flox/+}$, $R26^{R/+}$; $Foxa2^{iCre\Delta neo/+}$, $\beta\text{-}catenin^{flox/+}$, $R26^{R/+}$; $Foxa2^{iCre\Delta neo/+}$, $\beta\text{-}catenin^{floxdel/flox}$, $R26^{R/+}$; $Foxa2^{iCre\Delta neo/+}$, $\beta\text{-}catenin^{floxdel/flox}$, $R26^{R/+}$; $Foxa2^{iCre\Delta neo/+}$, $\beta\text{-}catenin^{floxdel/flox}$, $R26^{R/+}$.

(Abbreviations: h: heart; li: liver; lb: limb buds; pa: pharyngeal arches; fb: forebrain; s: somites; g: gut)

At P5 mice were dissected and genotyped (for distribution see chart 3). There was no obvious phenotype regarding organ organization or relative size. The staining pattern for β-galactosidase showed no obvious abnormalities either (see figure 31). But the mice indeed appeared a bit smaller than their littermates, yet this is also true for mice carrying the $\beta\text{-}catenin^{floxdel}$ allele alone. There is evidence that the Mendelian distribution is not fulfilled (see chart 3b, red asterisk) as it is also the case for the $\beta\text{-}catenin^{floxdel}$ allele alone (see chart 3a, red asterisk).
These results suggest that a deletion of one allele of β-catenin already leads to a disruption of the expected Mendelian ratio and that this effect is amplified if the second allele is also deleted under the control of Foxa2-iCre recombinase.

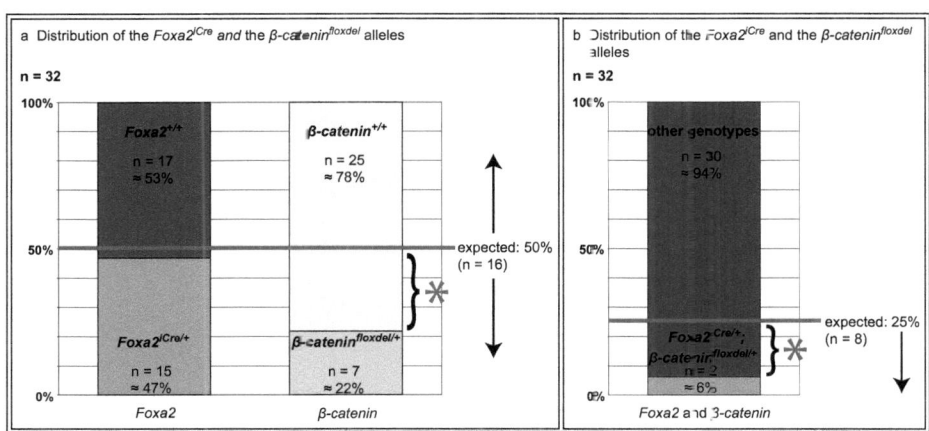

Chart 3: Mendelian distribution of $Foxa2^{iCre/+}$, $\beta\text{-}catenin^{floxdel/flox}$, $R26^{R/+}$ mice

The chart shows preliminary data of the Mendelian distribution of $Foxa2^{iCre/+}$; $\beta\text{-}catenin^{floxdel/flox}$; $R26^{R/+}$ mice. Because of the design of the mating all mice carry one $\beta\text{-}catenin^{flox}$ and one $R26^R$ allele. (n = 32)
Specified are the total number of mice carrying the $Foxa2^{Cre}$ versus the wild type alleles (a; left column), the $\beta\text{-}catenin^{floxdel}$ versus the wild type alleles (a; right column), and both alleles together versus all possible other genotypes (b). Percentages are designated within the columns and the expected ratio (calculated mendelian distribution) is marked with a red line in both a and b. Asterisks in a and b point out the differences of actual distribution of genotypes compared to the expected Mendelian distribution.

Results

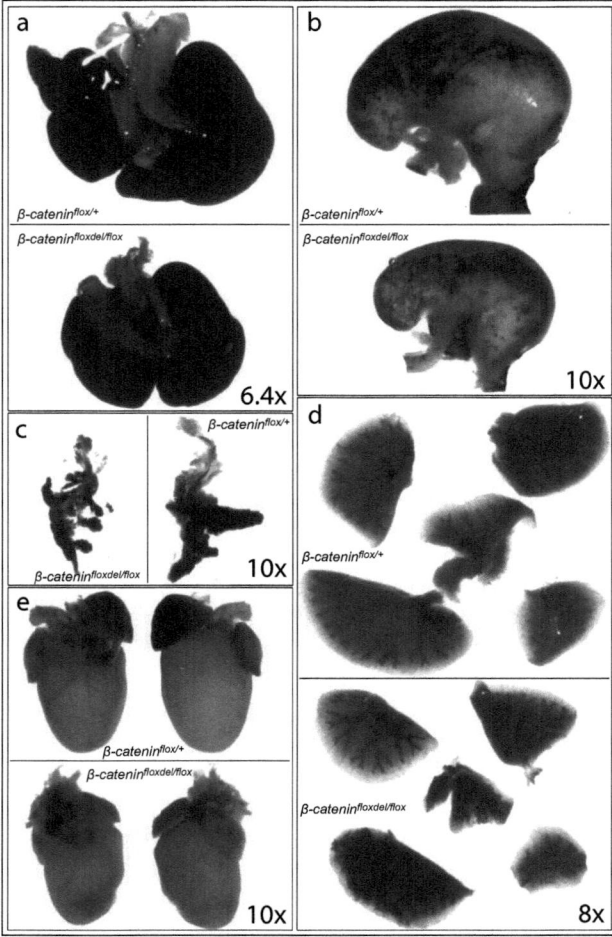

Figure 31: Organs of $Foxa2^{iCre/+}$; $β\text{-}catenin^{floxdel/flox}$; $R26^{R/+}$ mice at P5

The figure shows organs of P5 old mice, namely liver (a), stomach (b), pancreas (c), lung (d) and heart (e). Both mice carry the $Foxa2^{iCre}$ and the $β\text{-}catenin^{flox}$ alleles. The $β\text{-}catenin^{floxdel}$ allele is inherited like indicated in the figure. The mouse carrying the both the $β\text{-}catenin^{flox}$ allele and the $β\text{-}catenin^{floxdel}$ allele was in general a bit smaller than the one with only the $β\text{-}catenin^{flox}$ allele and consequently all its organ.

4.5. Differentiation of ES cells into endoderm

The obvious discrepancy between supply and demand of donor organs for transplantation clearly demonstrates that there is an increasing need for alternatives. One alternative holding many advantages (e.g. regarding immune reactions, availability of transplantable material etc.) could be the cell replacement therapy using *in vitro* differentiated (embryonic) stem cells. Suitable for cell replacement therapies are organs like liver and pancreas, or more precisely, pancreatic islet cells which are composed of mainly insulin-producing β-cells. At the current time, these organs are more promising for cell therapies than complex organs like the heart or the kidneys, where transplantation of single cells may not lead to sufficient therapeutic effects.

The first steps in the direction of cell replacement therapies are already made. There is one example where human ES cells were successfully differentiated *in vitro* into β-cells that had the ability to cure insulin insufficient mice after transplantation (Kroon *et al.*, 2008; D'Amour *et al.*, 2006) by blood sugar dependent production of insulin (for differentiation protocol see figure 32). These two different protocols available for the differentiation of ES cells into pancreatic endoderm precursors notably start with the same condition: Wnt and activin A conditioned medium without serum. However, one should be aware of the fact that the insulin-producing cells generated in this *in vitro* approach are clearly not comparable to adult β-cells, but rather have embryonic characteristics. After transplantation into a recipient mouse the human cells that were generated *in vitro* differentiated and matured for 30 days to three months until they reached their maximum of insulin secretion and showed glucose sensitivity (Kroon *et al.*, 2008).

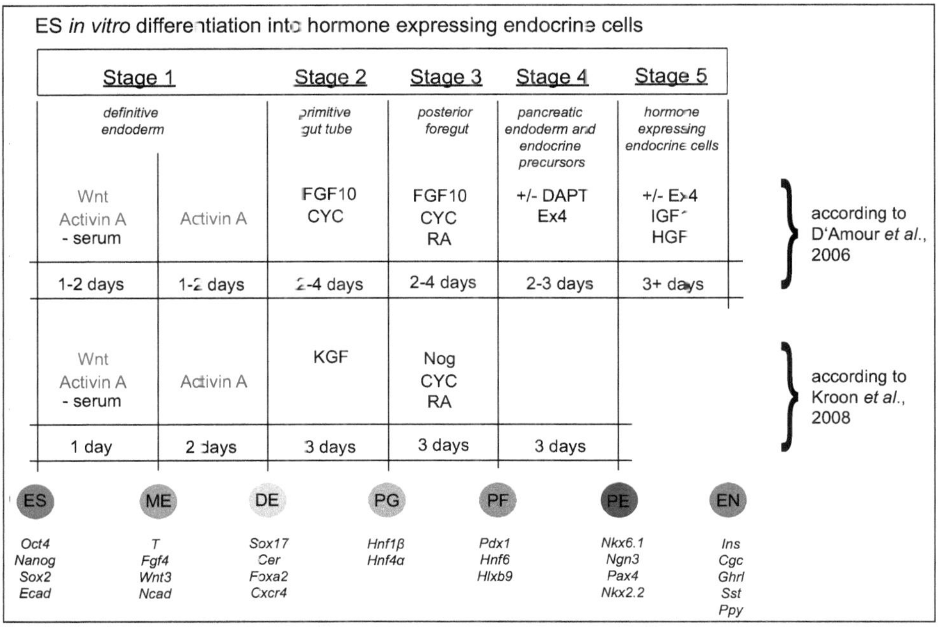

ES *in vitro* differentiation into hormone expressing endocrine cells

	Stage 1	Stage 2	Stage 3	Stage 4	Stage 5	
	definitive endoderm	primitive gut tube	posterior foregut	pancreatic endoderm and endocrine precursors	hormone expressing endocrine cells	
	Wnt Activin A - serum	Activin A	FGF10 CYC	FGF10 CYC RA	+/- DAPT Ex4	+/- Ex4 IGF1 HGF	according to D'Amour *et al.*, 2006
	1-2 days	1-2 days	2-4 days	2-4 days	2-3 days	3+ days	
	Wnt Activin A - serum	Activin A	KGF	Nog CYC RA			according to Kroon *et al.*, 2008
	1 day	2 days	3 days	3 days	3 days		

ES	ME	DE	PG	PF	PE	EN
Oct4 Nanog Sox2 Ecad	T Fgf4 Wnt3 Ncad	Sox17 Cer Foxa2 Cxcr4	Hnf1β Hnf4α	Pdx1 Hnf6 Hlxb9	Nkx6.1 Ngn3 Pax4 Nkx2.2	Ins Cgc Ghrl Sst Ppy

Figure 32: Protocol for differentiation of ES cells into blood sugar dependent insulin-producing cells
In vitro differentiation according to D'Amour *et al.* (2006) and Kroon *et al.* (2008). Specified are the factors added to the medium the time of differentiation under each single condition and the stages of pancreatic development that were achieved (Stage 1-5).
(Abbreviations: CYC = KAAD-cyclopamine; Cer = Cerberus 1 homolog; Cxcr4 = Chemokine (C-X-C motif) receptor 4; DAPT = γ-secretase inhibitor; DE = definitive endoderm; Ecad = E-cadherin; EN = hormone-expressing endocrine cells; ES = hES cell; Ex4 = exendin-4; Fgf4 = Fibroblast growth factor 4; Foxa2 = Forkhead box protein a2; Ghrl = ghrelin/obestatin prepropeptide; Hlxb9 = Homeobox gene HB9; Hnf1β/ 6/ 4α = Hepatocyte nuclear factor 1β/ 6/ 4α; Ins = Insulin; ME = mesendoderm; Ncad = N-cadherin; Ngn3 = Neurogenin 3; Nkx2.2/ 6.1 = Homeobox protein Nkx-2.2/ 6.1; Oct4 = Octamer-binding transcription factor 4; Pax4 = paired box gene 4; Pdx1 = Pancreas/duodenum homeobox 1; PE = pancreatic endoderm and endocrine precursor; PF = posterior foregut endoderm; PG = primitive gut tube; Ppy = pancreatic polypeptide; RA = all-trans retinoic acid; Sox2/ 17 = SRY-related HMG-box transcription factor 2/ 17; Sst = Somatostatin; T = Brachyury)

Regarding cell replacement therapies, the creation of iPS cells led to a new era in stem cell research (Takahashi et al., 2007; Takahashi and Yamanaka, 2006; Nakagawa et al., 2008). Induced stem cells could make it possible to obtain autologous differentiation systems where donor and recipient are the same person. It could be possible to first de-differentiate someone's cells to an ES cell-like state and then to differentiate those in a directed manner to the specific cells needed. These cells could than be transferred back to the donor (now operating as the receptor) and should not lead to any immune reaction because of the autologous nature of the tissue (especially in respect to identical major histocompatibility complexes I and II).

Taken together, establishing and optimizing in vitro differentiation systems is a fundamental goal to generate sufficient enough transplantable cell types in culture from ES or iPS cells.

4.5.1. Establishment of an in vitro differentiation system from ES cells into endoderm

For the study of the differentiation of ES cells into endoderm a protocol published by Tada et al. (2005) was adopted in the lab (also see Yasunaga et al., 2005). Different knock-in ES cells were therefore differentiated on collagen IV coated dishes in SFO3 medium supplemented with 10ng/ml human activin A and 0.1mM β-mercaptoethanol (see figure 33 and Material and Methods).

For differentiation T::GFP cells (Fehling et al., 2003) and Foxa2-Venus-fusion and Sox17-Venus- or Cherry-fusion cells were used. T::GFP cells express GFP under the transcriptional control of the mesendodermal and mesodermal marker gene Brachyury (T). The expression of the fluorescent protein therefore marks cells differentiated into mesendoderm and mesoderm (Fehling et al., 2003). Foxa2- and Sox17-fusion ES cell lines express the fluorescent proteins Venus (yellow) and or or Cherry (red) as fusions to the natural proteins and were generated by Dr. Ingo Burtscher by exchanging the translational stop-codon with the open reading frame of one of the fluorescent proteins. The complete 3'-untranslated region (UTR) is preserved this way and the transcriptional and translational regulation should therefore be as physiological as possible. The expression of the fluorescent protein in those cell lines can be used as an indicator for endodermal differentiation.

On protein level, neither Brachyury, nor Foxa2 or Sox17 are expressed in ES cells.

With the published protocol (Tada et al., 2005) a differentiation of all three cells lines was achieved, and was monitored by the expression of the fluorescent proteins (see figure 33).

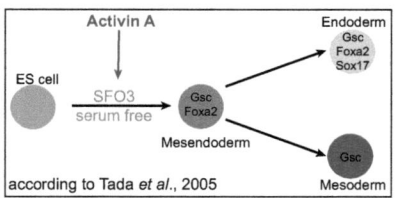

Figure 33: ES differentiation according to Tada et al. (2005)

The figure schematically shows the in vitro differentiation of ES cells into endoderm according to a protocol from Tada et al. (2005). ES cells (green) are treated with (human) activin A under serum-free conditions in SFO3 medium. They then start differentiating into mesoderm (red) and endoderm (yellow) proven by the expression of different makers (Foxa2, Sox17, Gsc) via a mesendodermal state (orange).
(Abbreviations: Foxa2 = Forkheadbox transcription factor a2; Gsc = Goosecoid; Sox17 = Sry-box related HMG box transcription factor 17)

To further optimize the differentiation of ES cells into Foxa2 and Sox17 expressing cells different co-culture systems were tested for efficiency and were microscopically compared to the differentiation on plain collagen IV coated dishes (for schematic protocol see 34a). Cell lines used for co-cultures were NIH3T3 either stably transfected with a vector expressing β-galactosidase (control system), Wnt1 or Wnt3a (Wnt1-NIH3T3, Wnt3a-NIH3T3). Figure 34b shows representative recordings of the differentiation efficiency. Already at day 1 of differentiation one can clearly observe that the differentiation using NIH3T3 cells expressing Wnt3a in co-culture is the most efficient, followed by Wnt1 expressing NIH3T3 (d1-2), control cells (d2) and the collagen coated plate without any co-culture (d3).

Furthermore, to show the difference in differentiation efficiency using Wnt3a over-expressing NIH3T3 compared to those, expressing β-galactosidase FACS analysis on Foxa2-Venus fusion differentiated cells was carried out over time. To be able to distinguish between ES cells and NIH3T3 cells even after differentiation, when the former ES cells cannot be as easily identified by their size and granularity (forward and side scatter), NIH3T3 cells were stably transfected with either a pCAGGS-Kozak-Tomato (ubiquitous red fluorescence driven by the chicken β-actin promoter) or a pCAGGS-H2B-Venus (nuclear yellow fluorescence driven by the same promoter)

construct. With the help of these constructs NIH3T3 cells became easily distinguishable due to their fluorescence no matter when or what fluorescence the ES cells would start to express during differentiation (for an example of Venus expressing differentiated ES cells cocultured on Wnt3a and Kozak-Tomato over-expressing NIH3T3 cells see figure 34d).
Figure 34c shows the expression of fluorescent fusion proteins in differentiating cells dependent on the coculture (+/- Wnt3a) used. The control FACS data for undifferentiated ES cells and Tomato over-expressing NIH3T3 cells are shown in figure 34c right box. The onset of Venus expression is visualized on the x-axis, the number of cells on the y-axis. As can clearly be seen the graphs for ES cells differentiated on Wnt3a over-expressing NIH3T3 cells (middle box in 34c) show a second Venus positive peak after 3 days of differentiation that grows over time (d3-d6), while the ES cells differentiated on β-galactosidase expressing NIH3T3 do not show any fluorescence in the period of differentiation that was focused on (left box in 34c).
Examples of what Foxa2-Venus fusion ES cells look like after differentiation on Wnt3a-NIH3T3 are shown in figure 34e-g'. The cells form networks (e and e') of tube-like structures (f and f') with thickenings and "holes" of fluorescence expression (g and g').
These results indicate that Wnt3a is necessary for an efficient differentiation of ES cells into endoderm and that also the feeder cells provide an enviroment with a positive impact on the differentiation.

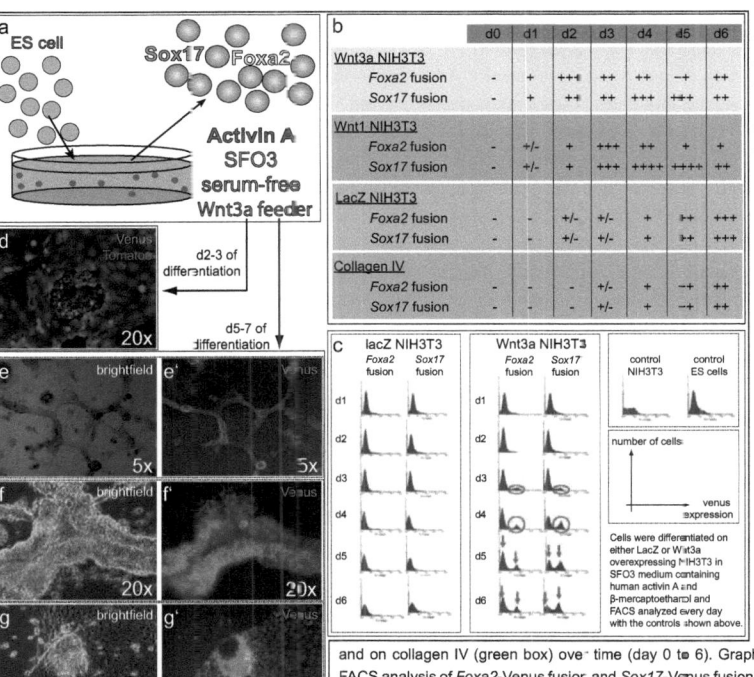

Figure 34: Optimization of differentiation using a co-culture system with NIH3T3 cells over-expressing Wnt3a

The figure schematically shows the protocol (a) for the optimized differentiation using a co-culture system with NIH3T3 Wnt3a over-expressing feeder cells (orange) in SFO3 medium (green) plus activin A (red). ES cells (light blue) differentiate into Foxa2 and Sox17 expressing endodermal precursor cells (orange and yellow). The table in b shows a comparison of the differentiations of Foxa2 and Sox17-Venus fusion ES cells on Wnt3a (yellow box), Wnt1 (pink box) and LacZ (blue box) over-expressing NIH3T3 cells and on collagen IV (green box) over time (day 0 to 6). Graphics in c show the data of FACS analysis of Foxa2-Venus fusion and Sox17-Venus fusion ES cells on Wnt3a (middle box) and β-galactosidase (left box) over-expressing NIH3T3 cells. Venus fluorescence is displayed on the x-axis while the number of cells is displayed on the y-axis at day 1 to 6 (d1-6) of differentiation. Fluorescent controls of NIH3T3 and undifferentiated ES cells are shown in the right upper box of c. As an example a Foxa2-Venus expressing colony of differentiated ES cells on Tomato and Wnt3a over-expressing NIH3T3 after 2-3 days is shown in d as an overlay of the two fluorescent pictures. In e-g examples of differentiations of Foxa2-Venus fusion ES cells for 5-7 days are shown in brightfield. The corresponding fluorescent pictures (Venus) are numbered e'-g'.
(Abbreviations: d = day; ES cells = embryonic stem cells; FACS = fluorescence activated cell sorting)

Results

4.5.2. Characterization of the *in vitro* differentiation system

To characterize the *in vitro* differentiated cells on a molecular level RT-PCR was performed on three independent ES cell clones with different differentiation status. RNA was then extracted at different time points of differentiation (d5 and d9) and the expression of marker genes was analysed. Differentiation efficiency could be monitored by the expression of fluorescent proteins under the control of the three early markers *T* (*Brachyury*, mes(end)oderm; T::GFP cells; Fehling *et al.*, 2003) *Foxa2* (mesendoderm; *Foxa2*-Venus fusion cells) and *Sox17* (endoderm; *Sox17*-Venus fusion cells).

Different primer sets were used in the semi-quantitative PCR to detect the expression of several markers (for primer sequences see Material and Methods), namely *Foxa2*, *Sox17*, *Pdx1* (*Pancreas/ duodenum homeobox gene 1*), IFAB-P (*intestinal fatty acid binding protein*), Cerl (*Cerberus like*), Hex (*Haematopoetically expressed homeobox gene*) and Nkx2.1 (*NK2 homeobox 1*). *Foxa2* and *Sox17* were used as controls of the differentiation into endoderm, as well as *Pdx1* as the first gene known to be expressed in the pancreas. IFAB-P marks the posterior gut endoderm, Cerl, as a Wnt inhibitor, is a marker for the more anterior endoderm, Hex marks the AVE as well as the earliest liver progenitors and Nkx2.1 the early progenitor of the lung. β-actin (not shown) was used as a control for the integrity and semi-quantitative concentration of the cDNA and reverse transcribed RNA extracted from an E9.5 embryo was used as a positive control for all markers. Wnt3a feeder cDNA and ES cell cDNA served as controls for the basic level of expression that can already be detected without differentiation.

RNA of many chosen markers is expressed in ES cells: *Foxa2*, *Sox17*, *Pdx1*, Cerl and Hex. *Sox17* and *Foxa2* were not detected in ES cells that carry a knock-in at the *Sox17* or *Foxa2* locus that replaces the Stop-codon with a fluorescent marker (unpublished observation). Additionally, the two factors could not be detected in immunostainings, suggesting that most of the transcription detected here is due to transcriptional noise in ES cells. Hex, however, seems to be expressed on a protein level since a BAC reporter line also shows an expression in the undifferentiated ES cell state (unpublished observation). The expression of the posterior marker *Sox17* seems to be upregulated between d5 and d9 of differentiation in T::GFP and *Foxa2*-Venus fusion cells or at least stays the same in *Sox17*-Venus fusion cells. The expression of *Foxa2* stays the same level in all three clones. The expression of the more anterior markers *Pdx1* (first pancreas marker), Cerl (Wnt inhibitor), Hex (first liver marker) and Nkx2.1 (first lung marker) are downregulated in this semi-quantitative assay (see figure 35).

The expression of the different markers, as measured by RT-PCR analysis, suggests that posterior endoderm is generated by using the *in vitro* differentiation assay.

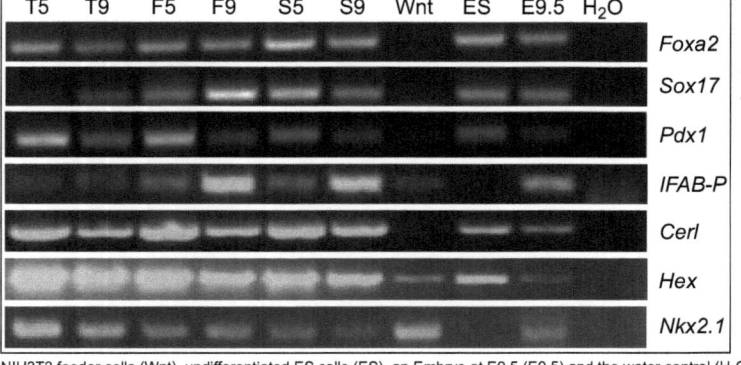

Figure 35:
RT-PCR on RNA samples of *in vitro* differentiated ES cells
The figure shows RT-PCRs for different markers (row 1-7: *Foxa2*, *Sox17*, *Pdx1*, IFAB-P, Cerl, Hex and Nkx2.1) on three different knock in ES cell lines (column 1-6: T = T::GFP, F = *Foxa2*-Venus fusion, S = *Sox17*-Venus fusion) under differentiation conditions after 5 (column 1, 3 and 5) and 9 (column 2, 4 and 6) days. Columns 7-10 show the controls: cDNA from Wnt3a NIH3T3 feeder cells (Wnt), undifferentiated ES cells (ES), an Embryo at E9.5 (E9.5) and the water control (H_2O).

4.5.3. *In vitro* differentiation – onset of marker expression (FACS analysis)

For differentiation of definitive endoderm *in vivo* the expression of *Foxa2* can be detected first (E6.5; Ang *et al.*, 1993), followed by the expression of *Sox17* (E7.0; Kanai-Azuma *et al.*, 2002). Resolution of static fluorescent microscopy and RNA expression analysis do not allow for a precise determination of the onset of marker gene expression *in vitro* for comparison. Other approaches allow for a more accurate detection of the fluorescence to distinguish the relative beginning of expression of the two markers.

To focus more precisely on the onset of marker expression and to be able to compare these data with what is known from the *in vivo* expression FACS analysis was used. In contrast to fluorescence microscopy FACS analysis is about 10-fold more sensitive. It is therefore possible to detect the onset of expression of fluorescent markers much earlier. *In vitro* differentiated ES cells either carrying knock-ins for *T* (GFP), *Foxa2* (Venus) or *Sox17* (Venus) – as mentioned above – were differentiated on Tomato-labelled Wnt3a overexpressing NIH3T3 cells. The FACS data of this differentiation over time clearly showed that the expression of GFP under the control of the *Brachyury* promoter can be detected first, followed by the expression of Venus under the control of the *Foxa2* promoter and Venus under the control of the *Sox17* promoter. While T:GFP expression is downregulated after 5 days of differentiation, Venus expression (under both – either or – regulatory elements of *Foxa2* and *Sox17*) stays constant (see chart 4). These data are in keeping with observations using fluorescent microscopy.

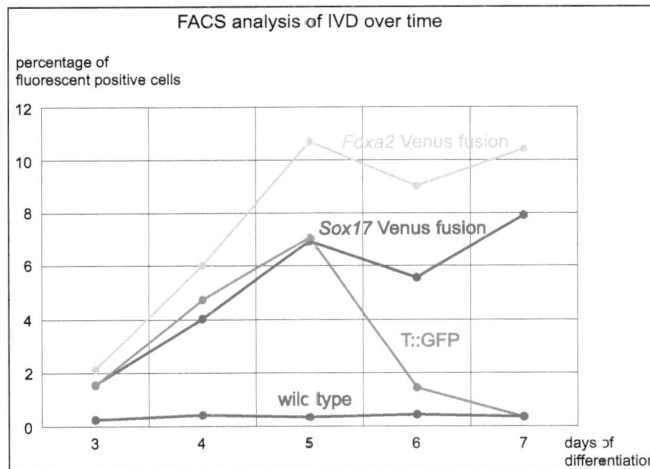

Chart 4:
FACS analysis of the onset of marker gene expression

The chart shows the expression of fluorescent markers under the transcriptional control of *Brachyury* (T; green), *Sox17* (red) and *Foxa2* (yellow) and their development over time (d3-d7). Stated are percentages of the cells that become fluorescent positive (y-axis) in relation to the differentiation time (x-axis). The gray line represents the negative control (undifferentiated ES cells without a fluorescent maker).

(Abbreviations: FACS = fluorecsence activated cell sorting; IVD = *in vitro* differentiation; *Foxa2*, *Sox17*, *T* = transcription factors, see text)

4.5.4. *In vitro* differentiation – onset of marker expression (live imaging)

To understand the kinetics and dynamics of Foxa2 and Sox17 expression in the *in vitro* assay, to identify potential endodermal subpopulations and to narrow down the time window of the onset of fluorescent fusion expression with cellular resolution, an ES cell line expressing both fluorescent proteins (*Foxa2*-Venus fusion and *Sox17*-Cherry fusion) was differentiated *in vitro* on unlabelled Wnt3a over-expressing NIH3T3 cells and live-imaged using a Leica confocal.

The movie was started after one day of differentiation when the first Venus positive cells were observable. The focus was put on different colonies that already showed some fluorescent positive cells and the movie was performed overnight. Cells of eight different colonies were counted for the onset of expression of the two fluorescent markers. Foxa2-Venus expression can be detected after 14.0h (+/-4.3h) while Sox17-Cherry expression visible after 24h (+/-4.7h). The results demonstrate that Sox17-Cherry can be detected about 9.6h +/-4.2h after the expression of Foxa2-Venus is detectable (see figure 36). It could never be detected before the onset of Foxa2-Venus fluorescence.

Taken together, in the *in vitro* differentiation assay *Foxa2* is expressed before *Sox17*. No subpopulation that is only *Sox17*-positive can be identified; *Foxa2* and *Sox17* are sequentially active in the same cells in a possible endoderm-specific molecular cascade with Foxa2 as a potential pioneering factor.

Results

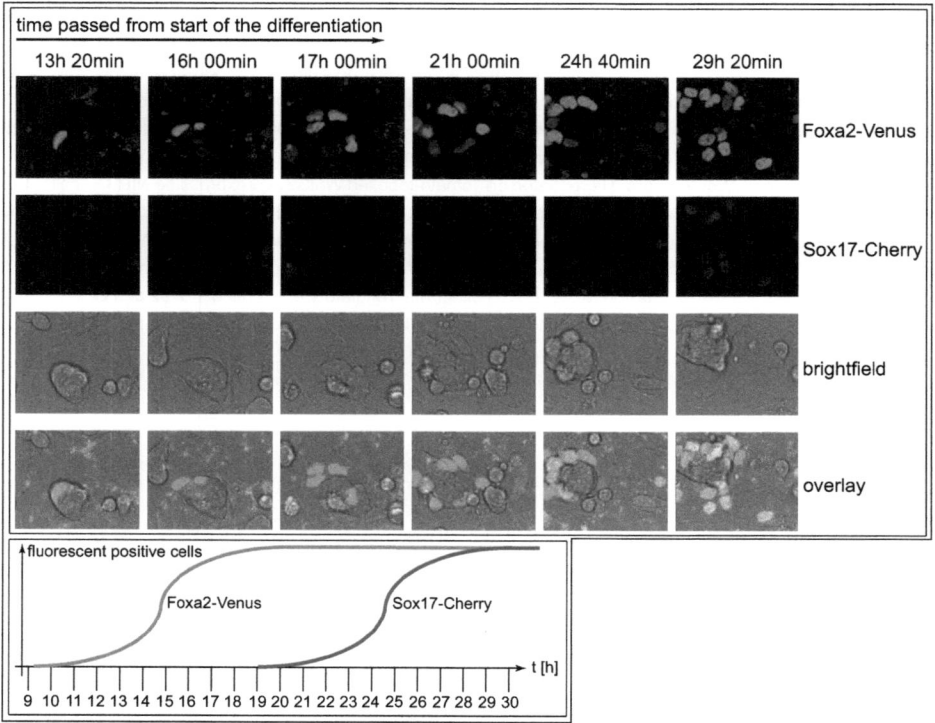

Figure 36: Live imaging of the onset of marker gene expression

The figure shows data from live imaging of an *in vitro* differentiation. In the first row pictures of the increasing Foxa2-Venus fluorescence of the cells over time are displayed (from left to right). The second row of pictures shows increasing Sox17-Cherry fluorescence over time in parallel to Foxa2-Venus fluorescence with a delay of several 9.6h (+/-4.2h). The third row shows brightfield pictures and the fourth row overlays from both fluorescent markers and brightfield.

Underneath the microscopic pictures a diagram elucidates the delay between Foxa2-Venus and Sox17-Cherry expression. The number of fluorescent cells is displayed on the y-axis, while the time of differentiation is displayed on the x-axis (not standardized).

4.6. Using the *in vitro* differentiation system to screen for micro RNAs influencing endoderm development

Micro RNAs (miRNAs) play important roles during embryonic development and in stem cell biology (for review see Cheng et al., 2005; Hatfield et al., 2005). Many of their functions have remained unravelled up until now. It is known that miRNAs regulate translation by binding to the 3'-UTR of their target mRNA thereby inhibiting its translation (Lee et al., 1993; Reinhart et al., 2000). Still it is unknown how target sites can be identified due to the fact that miRNAs do not show a 100% complementary sequence to their targets and also the "seed" region whose sequence was thought to have a high impact on binding to target mRNAs is not 100% complementary in several miRNAs (Didiano and Hobert, 2006). Bioinformatics have designed several tools for target prediction of miRNAs with about 300 target mRNAs for each miRNA (Brennecke et al., 2005; Bartel, 2004). Experimental approaches will be necessary to identify the real targets from a pool of computational predictions and to investigate their role in development, stem cell biology, metabolism and cancer.

With the knowledge about how miRNAs post-transcriptionally regulate expression and the tool generated with the fluorescent fusion ES cell lines the idea came up to use the *in vitro* differentiation system as a test system for miRNAs.

The *in vitro* differentiation allows for the induction of *Foxa2* and *Sox17* monitored by their fluorescent fusions knock-ins. The fluorescent fusions were thought to be useful as a read out for miRNAs. In principle, the prediction was that the fluorescent fusion protein is down-regulated in the same manner the wild type protein is when a miRNA that targets either *Foxa2* or *Sox17* mRNA or both is (over-) expressed, because the fusion and wild type mRNAs share an identical 3'-UTR and therefore should be controlled by the same regulatory mechanisms regarding miRNA activity (see figure 37).

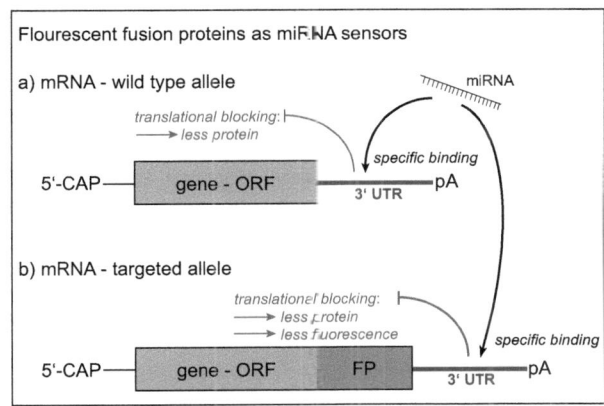

Figure 37:
Fluorescent fusion proteins as a read out of miRNA activity

The figure illustrates the prediction how the mRNA of the gene of interest (a) and its corresponding fluorescent fusion counterpart (b) are targeted by a specific miRNA. The miRNA specifically binds to the identical 3' UTRs of the two mRNAs and thereby leads to transcriptional inhibition of both. Inhibition of the translation consequently leads to less protein and therefore to less fluorescent protein and signal.
(Abbreviations: CAP = post-translational structure of mRNAs needed for the initiation of the translational complex; structure FP = fluorescent protein; miRNA = micro RNA, CRF = open reading frame, protein coding sequence; pA = polyadenylation signal; UTR = untranslated region)

4.6.1. Generation of the miRNA expression vector

The investigate the role of miRNAs during stem cell differentiation and their influence on the regulatory networks that are involved in that the combination of fluorescent fusion proteins and the *in vitro* differentiation system should be tested for its usability to unravel miRNA function in the molecular program of endoderm differentiation *in vitro*.

For this purpose naturally existing miRNAs were subcloned into a published expression vector, pUI4-SIBR (modified from Chung *et al.*, 2006). In this case a ubiquitously expressed polymerase II promoter (human ubiquitin C promoter) drives the expression of an artificially composed gene that combines an expression cassette for miRNAs (SIBR cassette = synthetic inhibitory BIC-derived RNA) and the open reading frame (ORF) of either GFP or puromycin resistance (see figure 38a). The expression of miRNAs using a RNA polymerase type II promoter has many advantages compared to the polymerase III promoter expression used for many shRNA expression vectors. Using a type III promoter does not allow for the expression of a marker gene with the same transcript. Additionally, polymerase III promoters do not allow for the expression of multiple RNAs. That means that multiple promoters or vectors would be required to knock down different genes (Yu *et al.*, 2003; Jazag *et al.*, 2005) or to express marker genes in parallel. Thus, miRNAs and marker genes might not be expressed at similar levels even if they are transferred by the same plasmid.

In the pUI4-SIBR plasmid the transcription of the SIBR-cassette containing the miRNA and a marker gene (either GFP or puromycin resistence) is driven by the same promoter. The expression of the marker gene is therefore directly correlated to the miRNA expression. The SIBR cassette is part of the gene *BIC*, a non-protein coding gene that carries an evolutionary conserved non-coding RNA involved in lymphoma and other types of cancer (for detailed information see Introduction, 2.5.1.; Clurman *et al.*, 1989; Eis *et al.*, 2005 van den Berg *et al.*, 2003 Kluiver *et al.*, 2005; Iorio *et al.*, 2005; Yanaihara *et al.*, 2006). Its third exon has been shown to code for miR155 (Lagos-Quintana *et al.*, 2002). BIC can be expressed from a heterologous RNA polymerase II promoter in a retroviral vector (Tam *et al.*, 2002). The smallest unit needed for the efficient expression of an artificial miRNA with the miR155 loop from the *BIC* locus is the SIBR cassette (Chung *et al.*, 2006). In the case of the pUI4 vector the expression unit (SIBR) is located within an intron allowing for the coupled expression of the miRNA and a marker

Results

gene (GFP or puromycin resistance; Chung et al., 2006).
Before the designed miRNAs were cloned into the pUI4 vector it was modified to have a resistance gene for selection in ES cells and a fluorescent marker as an indicator of different levels of expression of miRNA in different transgenic clones. The ORF of GFP was exchanged with the ORF of *H2B-CFP* (<u>c</u>yan <u>f</u>luorescent <u>p</u>rotein with a <u>h</u>istone <u>2B</u> nuclear localization signal) followed by an <u>i</u>nternal <u>r</u>ibosomal <u>e</u>ntry <u>s</u>ite (IRES) and a *puromycin* resistance gene (see figure 38a). After confirmation the vector was transfected into HEK293T cells to ensure functionality of the inserted cassette. Fluorescence with the correct localization and expression of the resistence gene were confirmed by fluorescent microscopy and the supplementation of the culture medium with puromycin, respectively (see Material and Methods).

4.6.2. Design of miRNA expression constructs

Using the generated vector for cloning of the miRNAs allows for the selection of clones in ES cells as well as for the qualitative detection of the expression of the miRNA and the transcriptional level of the miRNA as indicated by the fluorescent protein production. The miRNAs were designed as oligos according to the published protocol using the single site *BglII* within the SIBR cassette for cloning (see figure 38b, green letters; Chung et al., 2006). To allow for the determination of the orientation of miRNA integration, restriction sites were introduced with oligonucleotides: an *EcoRI* site 5' of the miRNA and a *BamHI* site 3'. These restriction sites are similar but not identical and allow for a not perfect binding within the miRNA stem after transcription as in natural miRNAs (see figure 38b, orange and turquoise letters). The natural loop of the miRNAs was exchanged with the loop of the miR155 used in the publication, because the loop is known to allow for the efficient processing of miRNAs (see figure 38b, red). Because it is also known that the distance between the first base pairing after the loop and the cleavage site is important for the cleavage activity of Drosha – and therefore the processing of the miRNA –, the sequence where Drosha is supposed to cleave within the natural miR155 was preserved (see figure 38, black letters; Chung et al., 2006; Zeng et al., 2005; Zeng and Cullen, 2005). The miRNAs were analyzed for predictive folding using the software "RNAshapes" that is available online. According to the predicted folding the complementary strand of the miRNA (see figure 38b, light blue) was altered if necessary for correct folding predictions, resulting in the presented miRNAs (see figure 39; Chung et al., 2006).

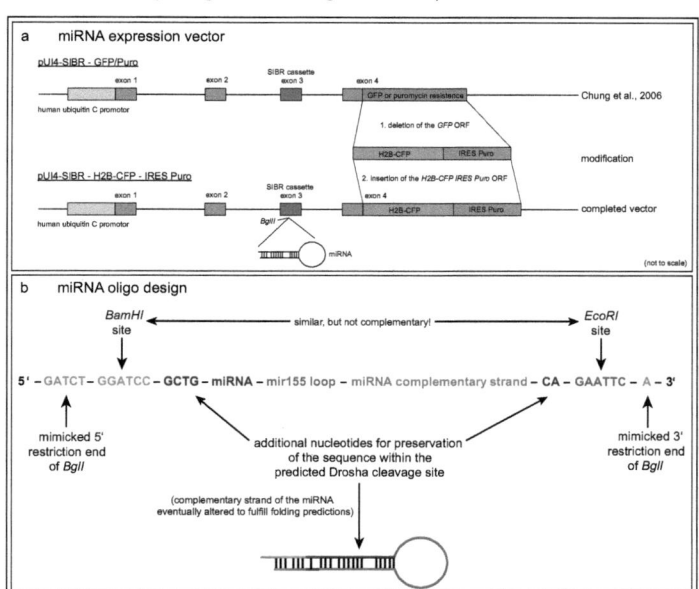

Figure 38:
Alteration of the expression vector for miRNAs and miRNA design

(a) The figure shows the modified miRNA expression vector. The ORF of GFP (green box) was exchanged with the ORF of *H2B-CFP* (blue box) plus a *puromycin* resistance attached to it using an *IRES* sequence (liliac box). For cloning of miRNA oligos into the SIBR cassette (red box) the *BglII* site can be used. The expression is driven by the polymerase II human ubiquitin C promoter (yellow box). The SIBR cassette is embedded into an artificial exon-intron structure (exons indicated as grey boxes).
(b) miRNAs oligos were designed to mimick a *BglII* overhang after cleavage at both 3' and 5' end (light green) when sense and antisense oligo bind complementary to each other. Furthermore they carry a *EcoRI* site 5' (orange) and a *BamHI*

58

site 3' (dark green) as indicated. The miR155 loop (red) is plotted and the additional nucleotides (black) used to preserve the cleavage site of Drosha are marked. Whenever required, the complementary strand of the miRNA was altered according to RNAshapes folding predictions. (Abbreviations: GFP = green fluorescent protein; H2B-CFP: cyan fluorescent protein linked to the histon 2B localization sequence; IRES: internal ribosomal entry site; miRNA: micro RNA; ORF: open reading frame; Puro = puromycin resistence gene; SIBR: synthetic inhibitory BIC-derived RNA)

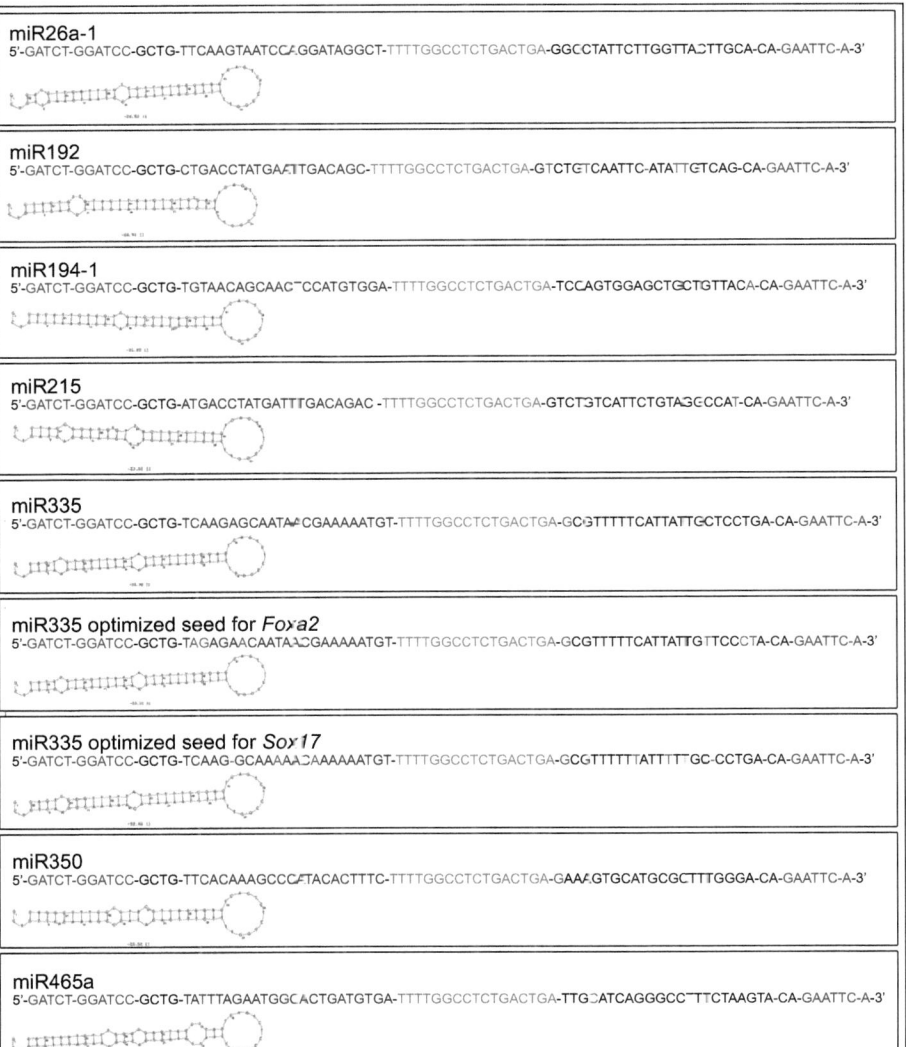

Figure 39: List of miRNAs used including folding predictions
Different miRNAs (1-9) and their predicted foldings according to the RNAshapes software. Alterations of the complementary miRNA strand that were made to preserve the miR155 loop structure according to RNAshapes are indicated in red. The loop is indicated in green, mimicked *BglII*, *EcoRV* and *BamHI* sites in blue, miRNA in grey.

Results

MiRNAs were selected for predicted impact on influencing the endoderm lineage. The general idea is that miRNAs and their target mRNAs are expressed exclusively, because miRNAs inhibit their target mRNAs from being translated into protein they downregulate the general noise of expression and thereby specifically suppress pathways (Stark *et al.*, 2005; Farh *et al.*, 2005). MiRNAs expressed in the endoderm should not suppress endodermal genes, but genes in other lineages (e.g. mesodermal genes) and vice versa.

Taking this aspect in account miR194, miR192 and miR215 were of special interest because they are known to be expressed exclusively in the endoderm in the developing Zebrafish and therefore likely support endoderm differentiation while suppressing other lineages (Wienholds *et al.*, 2003).

MiR335 was predicted to affect *Foxa2* and *Sox17* mRNA using several tools for the bioinformatic target analysis by Dominik Lutter (unpublished data). Additional information revealed that miR335 is coded within the gene *Mest*, a mesoderm specifically expressed factor, which further validates it as a candidate for the miRNA assay. As control experiments miR335 was additionally altered twice to optimize its sequence to the predicted target sequences of *Foxa2* and *Sox17* (see figure 39; *Sox17*- and *Foxa2*-optimized). MiR26a, miR465 and miR350 are miRNAs that were predicted to target *Sox17* (miR26a) and *Foxa2* (miR465a and miR350) using the same bioinformatic tools and that were used as controls for single knock downs of either *Foxa2* or *Sox17*. MiR26a and miR350 were already shown to be expressed in embryonic development at the same time the two endoderm differentiation factors, *Foxa2* and *Sox17*, are expressed.

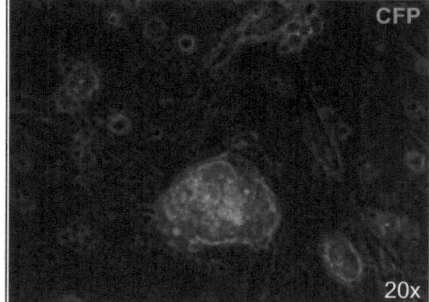

Figure 40: miRNA transgenic ES cell clone (example)
An undifferentiated ES cell clone that is transgenic for the pUI4-SIBR-H2B-CFP IRES-Puro miR26a. Expression of the genomic locus is visualized using CFP-fluorescence analysis. The pictures shows the overlay of brightfield and CFP channel.
(Abbreviations: CFP = cyan fluorescent protein; miRNA = micro RNA)

4.6.3. Fluorescent analysis of the miRNA-transgenic ES cell clones using *in vitro* differentiation and immunostaining

Stable ES cell clones were generated (see figure 40 and Material and Methods, 6.4.3. III.) and those with the highest expression of CFP, and consequently of miRNA, were differentiated using the established *in vitro* differentiation system. 24-well glass-bottom plates were used for differentiation to be able to differentiate 24 independent clones transgenic for the 9 different miRNAs. Fluorescence was checked every day under the same conditions for the control of the differentiation efficiency at the Zeiss inverse fluorescence microscope (AxioVert 200M, Zeiss). Thereby no effect on Sox17-Cherry fluorescence could be detected (see figure 41), except for miR335 *Sox17*-optimized. As shown in figure 41 a few colonies could be found that did not show any detectable nuclear CFP fluorescence. These colonies were positive for Sox17-Cherry. Additionally, most of the colonies showed a high CFP expression, but did not show any or only low Sox17-Cherry expression (see figure 41).

The analysis of fluorescence with the inverted fluorescent microscope showed no alteration in Sox17-Cherry expression for most of the clones.

Results

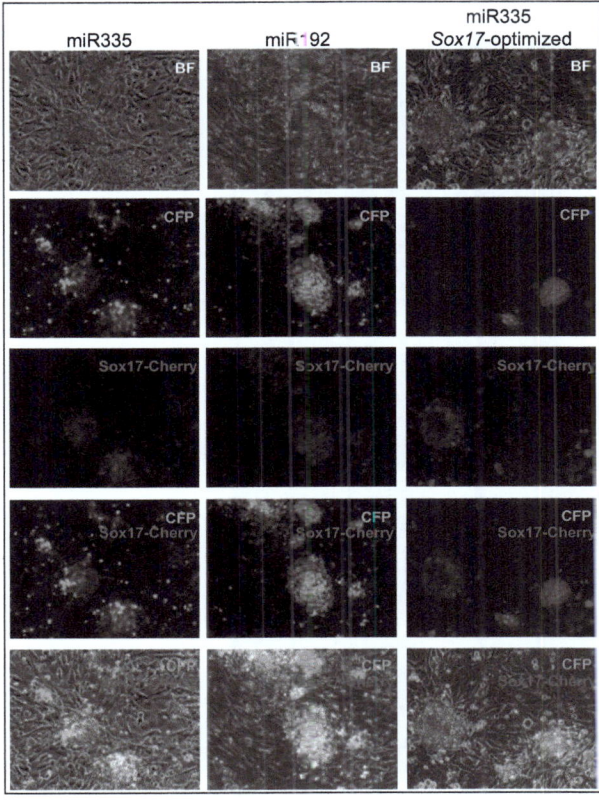

Figure 41:
In vitro differentiation of miRNA transgenic ES cell clones

Examples of Sox17-Cherry expression in miRNA transgenic *Sox17*-Cherry fusion ES cell clones at d4 of differentiation (20x). The first row shows the brightfield images, the second row the expression of the miRNA measured by the fluorescence intensity of H2B-CFP (blue), the third row shows Sox17-Cherry expression (red). The last two rows show the overlays of the two fluorescent channels and all three channels. Pictures in column 1 refer to miR335 clone 5, those in column 2 refer to miR192 clone 3, and in the last column the pictures refer to miR335 *Sox17*-optimized clone 6 transgenic *Sox17*-Cherry fusion differentiated ES cells.
(Abbreviations: BF = brightfield; CFP = cyan fluorescent protein; miR = micro RNA)

To analyse the effect of miRNA overexpression on endogenous *Foxa2* expression the cells were fixed after 6 days of differentiation and stained for Foxa2 protein (see figure 42a and 42b). Pictures were taken at the fluorescent microscope to compare differences in fluorescent levels. Therefore settings were kept constant for all pictures taken for each miRNA clone at each position (inverted fluorescent microscope AxioVert 200M, Zeiss).

For miRNA 192, 26a, 215, 350, 465a, 335 *Sox17*-optimized and 335 *Foxa2*-optimized, no alteration in the fluorescence intensity could be detected (see figure 42a and 42b). For miR335 a decrease of Foxa2 could be visually detected in two independent clones relatively to miR335 *Sox17*-optimized. MiR194 on the other hand showed a very strong and intensive expression of Foxa2 for two independent miR194 transgenic ES cell clones in comparison to miR335 *Sox17*-optimized ES cell clones. Consequently, the difference of the Foxa2 expression on a fluorescence level between miR194 and miR335 was even higher

In none of the other miRNAs an effect could be detected with the sensitivity of the fluorescent analysis at the inverted microscope, indicating that the effects that miR194 and miR335 show are specific and that they are not a result of clonal variation as also indicated by the two independent clones showing the same result.

Results

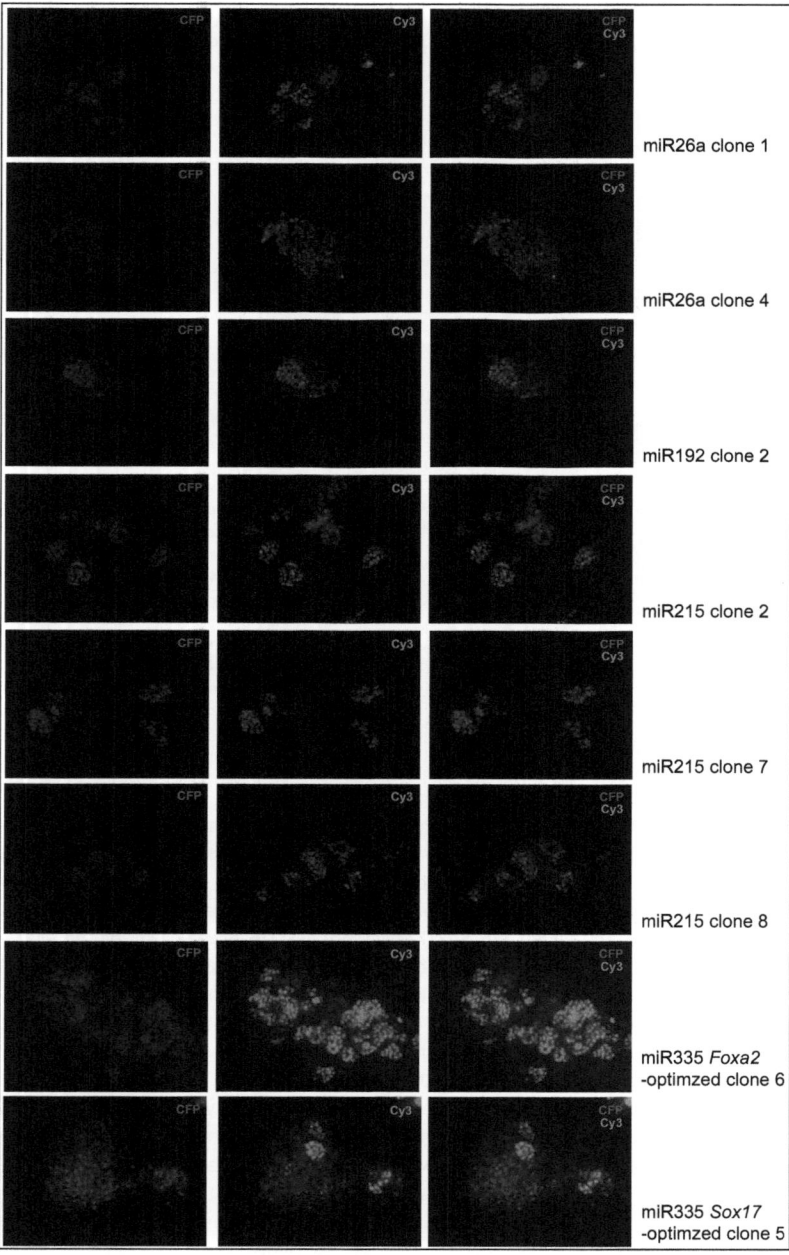

Figure 42a: Foxa2 immunostaining on differentiated miRNA transgenic ES cell clones at d6 of differentiation
(for figure legend see figure 42b)

Results

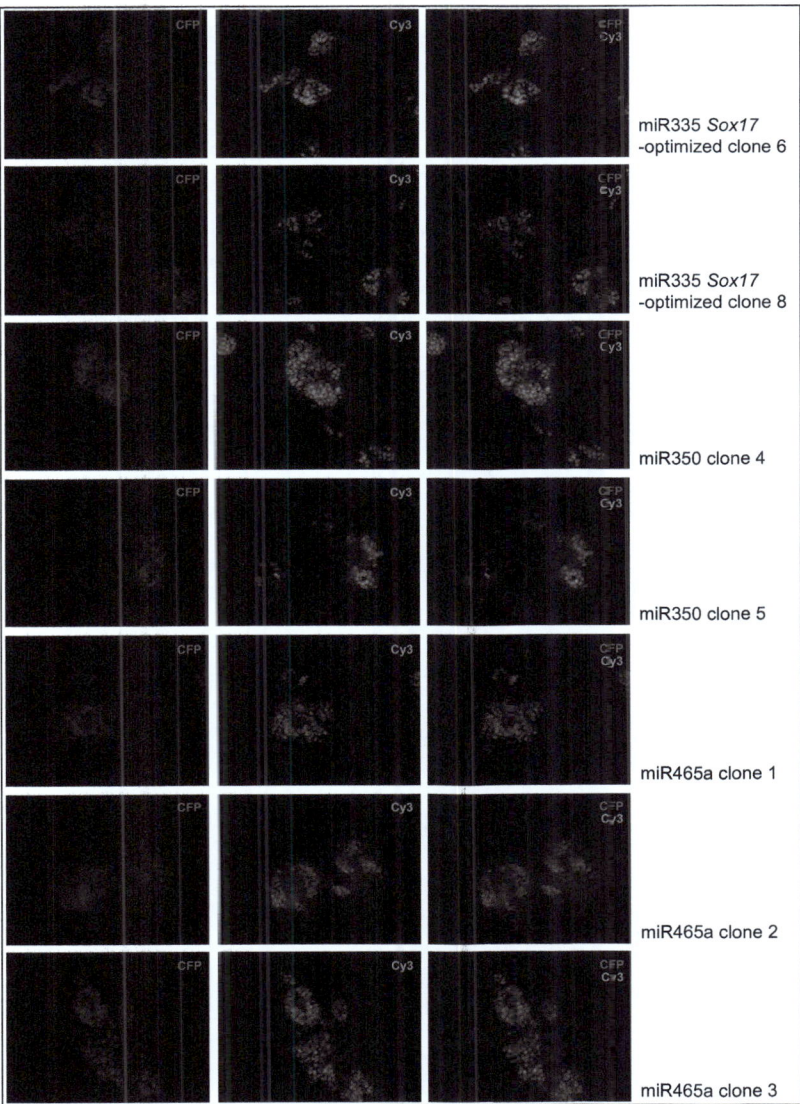

Figure 42b: Foxa2 immunostaining on differentiated miRNA transgenic ES cell clones at d6 of differentiation
For all different miRNAs constructs 1-3 different ES cell clones were differentiated *in vitro* and stained for Foxa2 (see right column for miRNA construct and clone number). In the first column the expression of the miRNA is shown measured by the fluorescence of H2B-CFP. In the second column the Foxa2 staining is displayed (Cy3 fluorescence) and in the third column the overlays of both CFP-fluorescence and Cy3-staining are presented. All pictures are taken with 20x magnification.
(Abbreviations: CFP = cyan fluorescent protein; ES cell = embryonic stem cell; H2B = histon 2B; miRNA = micro RNA)

Results

Statistical analysis with the help of binary pictures could verify the difference with statistical significance (for examples of binary pictures see figure 43). To achieve binary pictures a threshold of fluorescence intensity was determined using Otsu's method (Otsu, 1979). Fluorescence signals higher than the threshold count as 1, those signals that are lower than the threshold count as 0. This way each of the pictures can be translated into a matrix. Adding all matrices for one miRNA clone results in a number that is a value for the number of cells that have a high signal for both the miRNA (CFP fluorescence) and Foxa2 (Cy3 immuno-labelling). The higher the value is, the closer is the relation between high miRNA expression and high Foxa2 expression. Consequently, a miRNA that negatively regulates Foxa2 would result in a relatively low value.

Calculating the fluorescence for five different positions for each miRNA clone resulted in 24161 on average for miR335 (relatively low value compared to miR335 Sox17-optimized and miR194), 34396 for miR335 Sox17-optimized (median value compared to miR335 and miR194) and 92592 for miR194 (relatively high value compared to miR335 Sox17-optimized and miR335). The relation between the values is as expected: $value_{miR335} < value_{miR335\ Sox17\text{-optimized}} < value_{miR194}$. A low value is achieved with the mesoderm-specific miR335, showing that Foxa2 is negatively regulated when miR335 is overexpressed. A high value for miR194 shows that Foxa2 expression and the differentiation into cells expressing Foxa2 are even supported by expression of this miRNA.

The statistical significance was checked using the Wilcoxon rank-sum test (Wilcoxon, 1945). The p-value estimated for miR335 vs. miR194 is 6.3880×10^{-04}, for miR335 vs. miR335 Sox17-optimized = 0.0265 and for miR194 vs. miR335 Sox17-optimized = 0.0011. Taken together, all p-values are between 6.3880×10^{-04} and 0.0265 and therefore below p=0.05 and statistically relevant.

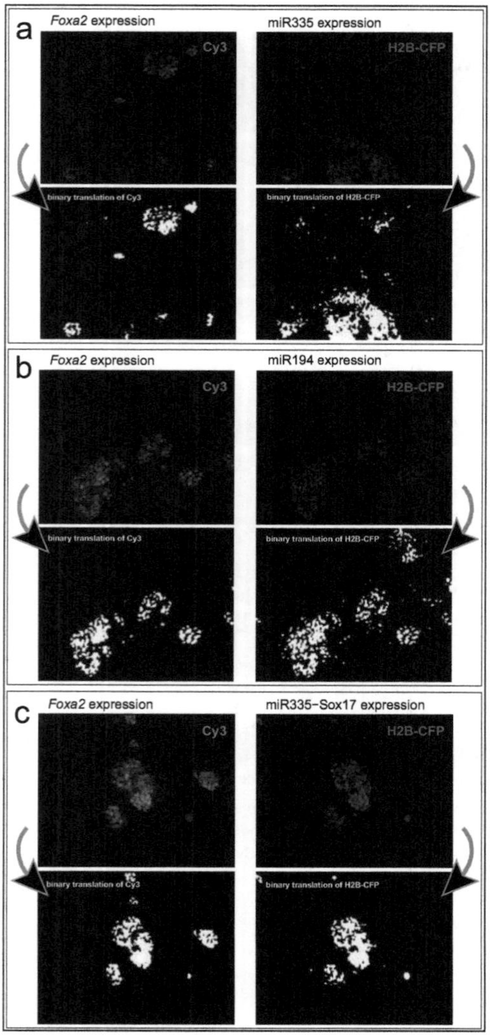

Figure 43:
Comparison of miR335, miR335 Sox17-optimized and miR194

At d6 of in vitro differentiation differentiated miRNA transgenic ES cells were stained for Foxa2 expression (displayed in green, Cy3 fluorescence). Fluorescent pictures were taken with the same settings for all positions. In a) Cy3 (left picture, Foxa2 staining, green) and H2B-CFP fluorescence (right picture, miRNA, red) of one position of the differentiated miR335 clone 8 is displayed. Below each fluorescent picture the translated binary picture is shown. Pictures for miR335 Sox17-optimized clone 6 (b) and for miR194 clone 3 are arranged in the same manner. Magnification is 20x. (Abbreviations: CFP = cyan fluorescent protein; ES cell = embryonic stem cell; H2B = histon 2B; miRNA = micro RNA)

In summary, it could be proven experimentally that miR335, coded within a mesoderm-specifically expressed gene, inhibits *Foxa2* expression (see figure 44), whereas the endoderm-specific miR194 supports Foxa2 expression for independent ES cell clones in the *in vitro* differentiation system. Both miRNAs might play important roles in embryonic development in the separation of mesoderm and endoderm (for expanded model of the differentiation of ES cells into endoderm see figure 44).

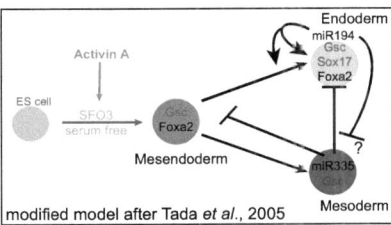

Figure 44:
Expanded model of endoderm differentiation *in vitro*:
A model of the suppression and support of the establishment of endoderm versus mesoderm by miR335

The figure illustrates a new expanded model of the *in vitro* differentiation of ES (green) cells into endoderm (yellow). MiR335 is most likely expressed in the mesoderm (red) where it directly suppresses *Foxa2* expression and possibly other factors necessary for endoderm development. MiR194 is expressed in the endoderm and positively regulates endoderm differentiation in an indirect manner by possible negative regulation of other lineages.

(Abbreviations: ES = embryonic stem; Foxa2 Gsc, *Sox17* = transcription factors, see text for further information)

5. Discussion

5.1. The generation of *iCre* mouse lines under the control of two endodermally expressed transcription factors

In the mouse the endoderm forms during gastrulation between E6.5 and E7.5. The two earliest marker genes known for the developing endoderm are the forkhead transcription factor *Foxa2* and the SRY-related HMG box transcription factor Sox17 (Monaghan et al., 1993; Sasaki et al., 1993). Deletion of either one leads to early embryonic lethality due to severe endodermal defects. *Foxa2* deficient embryos do not develop a distinct node or notochord; they have secondary defects in the organisation of the somites and the neural tube and do not form fore- and midgut, while *Sox17* knock-out embryos do not form mid- and hindgut (Kanai-Azuma et al., 2002; Ang and Rossant, 1994; Weinstein et al. 1994). The expression of *Foxa2* in the embryonic portion of the developing embryo starts at E6.5 at the posterior side in the region of the primitive streak. It is not only expressed in cells that form the endoderm but also in cell populations with important organizer function in the gastrula embryo, including the node, and in the notochord the organizer of the neural tube as well as in the floorplate. *Sox17* expression in the embryo proper starts briefly after *Foxa2* at E7.0 in the region of the primitive streak where the endoderm forms. It is still not known what potential *Foxa2*- and *Sox17*-positive progenitors have, what lineages they are able to generate and to what cell types and organs these cells will give rise to in the adult organism. Moreover, only little is known about the signals factors and pathways involved in the specification of certain lineages in the endoderm.

The aim of this thesis was to generate two Cre mouse lines which express Cre in endoderm progenitors under the control of *Foxa2* and *Sox17*, transcription factors that are essential for endoderm development. One intention was to lineage trace cells expressing the two transcriptions factors in the early embryo in order to identify the endodermal organs derived from cells that at one stage of development express one of the two transcription factors. Moreover, it should also allow for the identification of different populations that are positive for both or only one of these markers.

Gaining knowledge on the progenitor cells and characterizing those in more detail would support a better understanding of endoderm formation and the formation of its derivatives.

To elucidate the developmental mechanisms, the signals and factors that trigger endoderm specification and differentiation was the second intention for the generation of Cre lines expressed in a spatial and temporal manner matching the expression of *Foxa2* and *Sox17*. These Cre lines are the basis for the analysis of pathways that might be involved in endoderm development. They allow for the conditional deletion of key factors within signalling pathways in the particular subpopulations that express Cre. This way signals and pathways necessary for the development of cell populations and individual organs can be unravelled and the knowledge will benefit the understanding of pathological failure and diseases and is crucial for progress in regenerative medicine.

The knock-in strategy for both Cre lines was planned to maintain all cis-regulatory elements, to avoid nonsense-mediated decay and reduced expression levels due to alternative splicing. Moreover, knocking in the Cre should generate null alleles to be able to analyze their function in early development.

Deletion of the complete ORF which is the usual way to produce null alleles was not possible for both genes because they both had exon-intron spanning protein-coding sequences that, referenced from Mouse Genome Informatics (http://www.informatics.jax.org/) and Ensembl Genome Browser (http://www.ensembl.org/index.html), contained highly conserved regions suggesting that these might be regulatory elements that should not be destroyed.

In the case of *Foxa2* which has three exons the ORF starts within exon 2 and continues in exon 3, the full length transcript was verified by 5'-RACE (Lai et al., 1991). The structure of Sox17 regarding exons and introns and the ORF is comparable to *Foxa2*: The ORF of the 5 exon long gene starts in exon 4 and proceeds in exon 5.

Several splice variants with biological evidence based on cDNA clones were indicated checking the Ensembl Genome Browser. Therefore it was impossible to use any downstream exon of exon 1 for the knock-in without interfering with possible alternative splicing. The last exon in both cases was not considered a possibility, because

a fusion-protein would have been made due to the fact that the ORF starts one exon before.
However, the decision to knock the Cre into exon 1 had another consequence. It is known that a eukaryotic control mechanism that inhibits the translation of mutated mRNAs into defective proteins can lead to mRNA decay. Protein complexes that mark mRNA exon-exon boundaries after splicing are the basis for the so-called nonsense-mediated mRNA decay (NMD). The mediators of NMD recognize premature stop-codons that appear more than 50-55bp upstream of the 3' most splicing protein complex marking the exon-exon junction. It was shown that mRNAs whose translations terminate less than 50-55bp upstream of this junction are immune to NMD (Maquat, 2002). To avoid the possibility of nonsense-mediated decay by alteration of the sequence of the first exon a minigene approach was used. The ORF of the Cre in exon 1 terminates directly in front of an artificial intron to accurately follow the requirement of a stop-codon placed within the 50-55bp range before the first exon-exon junction.

Additionally, it was decided to use an improved Cre for both knock-ins instead of the commonly used Cre. The Cre (Sternberg and Hamilton, 1981; Sternberg et al., 1981), as a protein whose DNA sequence was derived from a bacteriophage, does not employ optimal codon-usage for eukaryotes and can be epigenetically silenced during development due to a variety of *CpG islands* that are sensitive to methylation and consequently mutations and epigenetic inactivation (Cohen-Tannoudji et al., 2000). Therefore a codon-improved Cre (iCre) that has the same sequence of amino acids, but a different genetic code was used to guarantee optimal conditions for Cre expression regarding the level of Cre and its consistency over generations (Shimshek et al., 2002). The iCre is improved for use in mammals regarding codon-usage preferences (Haas et al., 1996) with the help of silent base mutations. Additionally it is upgraded for translational efficiency with a perfect *Kozak* consensus sequence at its 5'-end (Kozak, 1997) and following the N-end rule for potentially increased protein stability it is advanced with an additional valine codon in front of the simian virus 40 large T nuclear localization signal (Varshavsky, 1997).

5.2. $Sox17^{iCre/+}$ mice – a valuable tool for specific deletion of genes in arteries and an indication of a second promoter activated in the endoderm

After generating the $Sox17^{iCre/+}$ mouse line the first questions to be answered were if the Cre activity would reflect the previously documented expression of *Sox17*, what organs the *Sox17*-positive progenitors will give rise to, and if a mouse homozygous for the Cre allele would be embryonic lethal as predicted.

For Cre expression analysis, $Sox17^{iCre/+}$ mice were crossed to ROSA LacZ reporter mice ($R26^{R/R}$) to visualize Cre recombination activity by X-Gal staining. Recombination activity of Cre leads to irreversible ubiquitous expression of β-galactosidase wherever Cre is expressed or was expressed, in other words, all Cre expressing cells and their progeny will stain because of β-galactosidase expression from the ROSA locus.

Unexpectedly, $Sox17^{iCre/+}$ mice hardly showed any recombination activity in the anterior definitive endoderm and the hindgut or derivatives of these tissues at E9.5 or at P1 as described (Kanai-Azuma et al., 2002). There was no staining detectable in the lung or in the gut, although *Sox17* is known to be expressed in the lung (Wan et al., 2004) and in the progenitors of the gut (Kanai-Azuma et al., 2002).

In concert with different publications (Matsui et al., 2006; Sakamoto et al., 2007) from E9.5 onwards expression could be detected in vessels, or more precisely in the endothelium lining the vessels and its progenitors. As it was later shown by Perry Liao (Liao et al., 2009, Genesis), the expression in the vessel endothelium is restricted to arteries with one exception: the umbilical vein. This vein is the temporal connection between the embryo and the placenta and transports oxygen-rich blood, like arteries. Additionally, Perry Liao could show that mice homozygous for the iCre knock-in are viable.

Using the Genomatix Software package a previously unknown fourth promoter region could be identified that is located in intron 3 (see figure 13), predicted by the ElDorado and Gene2Promotor annotation software ("PromotorInspector", Eldorado/ Gene2Promotor Release 4.6, Genomatix Software GmbH). It is supported by 71 Cap Analysis Gene Expression (CAGE) tags in comparison to the first three promoters identified with 5 and 69 CAGE tags, respectively (Shiraki et al., 2003). The CAGE method allows for the identification of sequence tags corresponding to 5' ends of mRNAs at the cap sites and the identification of start points of transcription. The

Discussion

CAGE tags identified an mRNA that completely covered the coding sequence of Sox17. From the alternative transcriptional start site in intron 3 upstream of the ORF starting in exon 4 an alternative mRNA of Sox17 is transcribed as also predicted by the Human And Vertebrate Analysis and Annotation (HAVANA) group of the Sanger Institute. Of course, the 5' UTR of the formerly known transcript and the one identified by CAGE and HAVANA differ in length and part of the sequence (see figure 13).

The most upstream promoters of an elongated exon 1 were also identified – although both of the first promoters should drive the expression of the iCre knocked into exon 1 (see figure 13). The third promoter region within exon 2 only has associated transcripts that are not overlapping with the ORF of Sox17.

The results from the Cre expression studies clearly show that the fourth promoter downstream of the previously identified two promoter regions around exon 1 drives the expression of a transcript that obviously leads to the production of the same protein in the endoderm. The functional proof for this is that homozygous $Sox17^{iCre/iCre}$ mice are viable and do not recapitulate the knock-out phenotype as one might expect (Kanai-Azuma et al., 2002). The fact that homozygous mice are viable although they show strong Cre activity restrictive to the arterial endothelium can be explained by the redundancy of Sox17 and Sox18 in the developing vasculature (Sakamoto et al., 2007); another possibility that, of course, cannot be excluded at this stage is that the fourth promoter is additionally active in these domains. A Sox17 immunostaining on homozygous Cre mutants could elucidate this unsolved question.

A closer look at the fourth promoter may be warranted to figure out if its activity is restricted to the endoderm or if it drives Sox17 expression in the endoderm as well as in the endothelium of the veins and possibly of the arteries. Although there was no detailed discrimination made in former publications (Matsu et al., 2006; Sakamoto et al., 2007), a recently generated mouse line allows for the conclusion that Sox17 is generally expressed all over the vasculature (Sox17-2A-iCre; unpublished observation). It is therefore likely that the expression in vein endothelial cells is driven by the downstream promoter. Transgenic mouse lines in which Cre is only driven by the fourth promoter might be interesting to generate and to study. It may be possible to create an endoderm-specific Sox17-Cre with this approach. These Cre lines would have the advantage that their restricted expression would make it possible to study conditional knock-outs only within the endodermal Sox17 expression domain which are not due to deletion of genes within the endothelium of the vasculature that could have an impact at later stages when the vasculature is established (E9.5 onwards). On the other hand it might also be interesting to see if the knock-out phenotype can be replicated by deletion of intron 3, the domain of the fourth promoter. With the knowledge of tissue-specifically active promoters it is possible to also identify tissue-specific transcriptional regulatory elements to unravel the signalling pathways upstream of Sox17 in the different lineages.

The analysis so far also suggests that the expression of Sox17 in arteries versus veins is regulated at least partially via different pathways. Because the endothelium of arteries and veins has to fulfil different tasks regarding its resistance towards blood pressure and composition (oxygen-rich vs. oxygen-poor blood) it could be interesting to have a closer look at the different regulative mechanisms to identify more artery or vein specific marker genes. Additionally, for studies of cancer growth and angiogenesis, it might be interesting to follow Sox17 expression or specifically delete genes in the $Sox17^{iCre}$ expression domain, those cells that give rise to arteries, to analyze the underlying interactions and downstream pathways. Transferring the knowledge gained from developing arteries in the embryo to cancer research might help to find ways to inhibit the connection of the tumor to the circulatory system and thereby its growth and metastasis. Moreover, anticipating the most likely situation that artery outgrowth is based on the same mechanisms in development and cancer, inducible Sox17-Cre lines that are generated using the same knock-in strategy could be a useful tool for cancer models. Being able to induce the recombination activity of the Cre at a certain time point would make it possible to use it at later stages when tumors arise.

At this point it should be mentioned that new data supplied by the ElDorado annotation software (Eldorado Release 4.5, Genomatix Software GmbH) identify a gene encoded on the complementary strand. The transcriptional start site is located within the elongated exon 1, so the knock-in of the iCre should not interfere with the ORF of the gene itself, although the promoter overlaps with our knock-in. A blast search of the genomic sequence identified it as a part an unknown gene (RIKEN clone: BB656573) with expression in E12 embryos in the spinal ganglion To counter the argument that the alteration within the promoter of the non-annotated gene has influence on

Discussion

necessary pathways it should be mentioned again that mice homozygous for the knock-in are viable and fertile and have no striking phenotype.

In summary, the $Sox17^{iCre/+}$ mouse line may be a useful tool for studying angiogenesis, possibly not only during embryonic development, but also in clinical cancer studies. Additionally the mouse line provides functional proof for another promoter active in the endoderm. The results of the expression analysis and the genomic predictions unravel new aspects of the differential genetic regulation of $Sox17$ in the endoderm and the endothelium of arteries and veins.

5.3. The $Foxa2^{iCre/+}$ mouse line – a new tool to analyze gene function in a variety of tissues

For the lineage tracing of $Foxa2$ positive progenitor cells to investigate what these cells give rise to in the adult organism the generated $Foxa2$-iCre mouse line was crossed to $R26^{R/R}$. This way Cre recombination activity can be visualized by X-gal staining for the enzymatic reaction of β-galactosidase conditionally expressed from the ROSA locus.

As shown for another $Foxa2$ Cre line (Frank et al., 2007) recombination in the visceral endoderm and the early epiblast could not be observed although in situ data clearly demonstrate that (at E6.5-7.0) $Foxa2$ mRNA expression can be detected in the extraembryonic visceral endoderm and the embryonic posterior epiblast, cells which are fate mapped to give rise to the anterior mesendoderm (Perea-Gomez et al., 1999; Kinder et al., 2001). A possible explanation for this phenomenon is that working with Cre can lead to a time shift of about 12 hours due to delays associated with the transcription, translation, and accumulation of the Cre, followed by recombination, transcription, translation and accumulation of β-galactosidase. The concentration of Cre might be to low to be detectable.

However, the recombination activity faithfully reflected the reported expression of $Foxa2$ in the node, notochord and floorplate, tissues with organizer function, as well as all analyzed endoderm derived organs in line with in situ expression data (Monaghan et al., 1993; Sasaki and Hogan, 1993; Ang and Rossant, 1994), suggesting that $Foxa2$ marks an early progenitor population of the endoderm lineage. The expression data from early embryonic stages until E9.5 are consistent with the results obtained from another $Foxa2$ Cre mouse line, $Foxa2^{iTA}$ (IRES tetracycline transactivator; Frank et al., 2007) and the $Foxa2^{mem}$ (a tetracycline inducible Foxa2 Cre mouse line; Park et al., 2008).

However, using the $Foxa2^{iCre/+}$ mouse line is clearly easier in respect to the accomplishment of the cell lineage tracing as timed and renewed tetracycline injections are not necessary for the activation of the Cre. Also the $Foxa2^{iCre/+}$ mouse line shows a broader expression domain than that reported for the $Foxa2$ NFP-Cre mouse line ($Foxa2$ notochord-floorplate enhancer-Cre; Kumar et al., 2007). The $Foxa2$ NFP-Cre mouse line only shows Cre recombination activity in the notochord, floorplate and in the hindgut. However, one must note that in this transgenic line the expression of the Cre recombinase is only driven by the notochord-floorplate enhancer (Sasaki and Hogan, 1996; Nishizaki et al., 2001; Kumar et al., 2007) and the weak basal heat shock promoter $Hspa1$ ($Hsp68/70$; Kothary et al., 1989). Inducible Cre lines like $Foxa2^{iTA}$ are interesting tool for conditional deletion of genes at later stages to study their impact on development at different points of lineage restriction, determination and differentiation (Frank et al., 2007). Transgenic Cre lines whose expression covers only part of the endogenous expression ($Foxa2$ NFP-Cre) can be also interesting in this respect. They allow for deletion of genes within distinct subpopulations of the endogenous expression domain (Kumar et al., 2007).

In the organs of $Foxa2^{iCre\Delta neo/+}$ mice at later stages (from E12.5 onwards) the staining was consistently high in pancreas and liver as well as in the epithelium of the stomach and throughout the gut. Lung, thyroid and thymus, however, showed consistent chimaeric expression, unexpectedly. Reported mRNA expression in the lung even at later stages (E14.5 onwards; Wan et al., 2004; unpublished in situ data from Moritz Gegg, Institute of Stem Cell Research, Helmholtz Zentrum München) should have led to full recombination in the lung epithelium at one point. Even after birth, although Foxa2 expression is necessary for the transition to air-breathing after birth by genetically controlling a pulmonary program necessary for lung epithelia cell maturation (Wan et al., 2004), the lung still showed a chimeric expression.

Expression in the epithelium lining the heart was not expected from the reported *in situ* data either and showed only low penetration. However, in the *Foxa2^mem* mouse line (Park et al., 2008) and in a newly generated *Foxa2*-iCre mouse line (*Foxa2*-2A-iCre, unpublished observation, Moritz Gegg) expression in the heart is shown even earlier, at E9.5. The cells that are positive for Cre activity in both cases might give rise to the cardiac epithelium at later embryonic stages and in the adult heart as in the case of *Foxa2^iCre/+* mice.

Because of the chimaeric expression in several tissues and the fact that the *Foxa2^iCre/iCre* mice were viable and did not show the predicted null mutation, further analysis of the genetic locus were carried out. These analyses revealed that there is a second promoter within the first intron downstream of the previously identified. The second promoter was predicted based on expression data of the Human And Vertebrate Analysis and Annotation (HAVANA) group of the Sanger Institute (see figure 24 and ENSEMBL) and the Genomatix software package (Eldorado Release 4.5, Genomatix Software GmbH). The promoter prediction is supported by 30 CAGE (<u>C</u>ap <u>A</u>nalysis <u>G</u>ene <u>E</u>xpression; Shiraki et al., 2003) tags, in other words based on 30 full-length cDNA clones (as explained before, see *Sox17^iCre* allele).

The second promoter does not only explain why the mutation in exon 1 does not reflect the knock-out situation, but could also be an explanation for the chimaeric recombination at E7.5 and later stages in tissues like the lung, the thymus and the thyroid. The delay that emerges because the Cre has to be transcribed, translated and reach a certain threshold to be active can only partially explain how the chimaerism arises. Even at E14.5, when there is strong expression of *Foxa2* detected in the lung (based on *in situ* data produced by Moritz Gegg), no further recombination could be observed. The second promoter might drive the expression of *Foxa2* in these tissues at a higher level than the first promoter does. Because a certain threshold in copy number has to be obtained to induce recombination activity of the Cre, organs like the lung might still show a chimaeric pattern (discussed later).

In conclusion, a *Foxa2^iCre/+* mouse line was developed, which is useful to analyze gene function in endoderm precursor cells as well as in cells of the endoderm-derived organs namely, liver, pancreas and GI tract. Interestingly, it could be shown that all analyzed endoderm-derived organs faithfully express β-galactosidase from the *R26* locus suggesting that *Foxa2* marks an early progenitor population of the endoderm lineage.

The data presented in this work support the idea that the hypothetical endodermal stem cell expresses *Foxa2*. Besides the fact that the lineage tracing of *Foxa2*-positive cells described here marks all analyzed endoderm-derived organs, *Foxa2* is known as a *pioneering factor* in a higher regulative order opening the chromatin for other transcription factors (Shim et al., 1998; Chaya et al., 2001; Cirillo et al., 2002). The knock-out of *Foxa2* that is not capable of developing any endoderm-derived organs (Ang and Rossant, 1994; Weinstein et al., 1994) is also evidence for the endoderm stem cell theory as well as the expression in the adult gut. The gut endothelium is replaced by self-renewal and differentiation of the crypt cells, an adult stem cell population. This turn-over of the receptive gut endothelium is illustrated in figure 45. Crypt cells are cells located at the bases of the crypts; they divide and give rise to more differentiated cells that replace the crypt endothelium continuously from the bottom to the tip of the crypts (for review see Crosnier et al., 2006). The lineage tracing revealed that all cells in the gut epithelium are *Foxa2^iCre* positive, suggesting that the crypt cells themselves are already expressing *Foxa2^iCre*. This hypothesis is also supported by first observations of the localization of fluorescence in the murine gut with the help of a mouse line expressing a fluorescent marker under the control of *Foxa2* (unpublished observation, Dr. Ingo Burtscher).

However, the final proof that the *Foxa2*-positive progenitor cells have stem cell potential can only be addressed by clonal analysis, showing that one *Foxa2*-positive cell is in principle capable of forming all endoderm-derived cell types (Rossi et al., 2006; Lawson et al., 1991). The fact that single cell labelling and tracking in the embryo is difficult compared to *in vitro* systems, this goal can be accomplished with live imaging of *in vitro* differentiations or a combination of both: *in vitro* differentiation of ubiquitously marked ES cells to *Foxa2*-positive progenitor cells followed by injections of these cells into blastula-stage mouse embryos to find out what cell types they can differentiate into. Besides, based on the definition of a stem cell, the requirement of self-maintenance *in vitro* would have to be shown.

Taken together, the *Foxa2^iCre/+* mouse line is also a valuable tool which can be used to analyze gene function in

Discussion

the node, notochord and floorplate of the neural tube: cell types with important organizer activity. Moreover, if the hypothesis of *Foxa2* expression in the crypt cells, the adult stem cells of the gut, holds true it is also possible to study conditional deletion of genes in the gut stem cell compartment and thereby unravel mechanisms, pathways and factors involved in their maintenance and differentiation.

The expression analysis presented here strongly supports the hypothesis that *Foxa2* might be a marker for the endodermal stem cell.

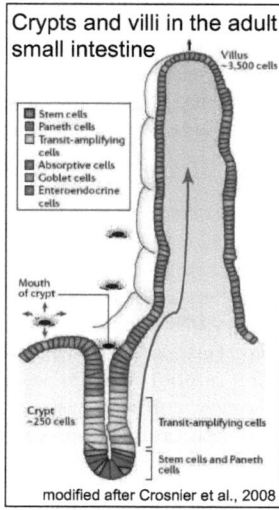

modified after Crosnier et al., 2008

Figure 45: Villus cell replacement in the adult small intestine
The figure shows a villus with one of the crypts that contribute to renewal of the villus epithelium. Black small arrows and the red arrow indicate the upwards flow of cells out of the crypts. Stem cells lie near the crypt base; it is uncertain whether they are mixed with, or just above, the terminally differentiated Paneth cells which secrete antibacterial proteins. Above the stem cells are so-called transit-amplifying cells (dividing progenitors, some of them already partially differentiated); and above these, in the neck of the crypt and on the villus, lie post-mitotic differentiated cells (absorptive cells, goblet cells for mucus secretion and enteroendocrine cells for secretion of different gut hormons). In the colon there are no villi, but the organization is otherwise similar; cells are discarded into the gut lumen after they emerge onto the exposed flat surfaces around the mouths of the crypts. (Modified after Crosnier et al., 2008)

5.4. The $Foxa2^{iCre}$ allele, a hypomorph with interesting potential

As $Foxa2^{iCre/iCre}$ mice were viable, a closer analysis of the mutant mice was undertaken.

Western blot experiment showed that there is indeed less Foxa2 protein produced in heterozygous and homozygous animals in the liver and in the pancreas (see figure 23 and 26b). Only minor differences of expression level could be found in the adult lung (see figure 26a). It is conceivable that a compensatory mechanism allows for a rise in the expression driven by the second promoter in order to normalize the Foxa2 protein level. Another possibility is that the first promoter under whose control the *iCre* is transcribed is only active during certain periods of development and within certain tissues (e.g. the liver) and that the other promoter is responsible for *Foxa2* expression the rest of the time.

Using RT-PCR, it could also be shown that iCre is not expressed in all tissues analyzed. At least at later stages the activation of the first and the second promoter is obviously tissue specific. E.g. the first promoter might be active in the lung precursor cells but the expression is either too low or is only there for a short period during development and therefore iCre cannot accumulate, leading to chimeric recombination. At later stages the second promoter may preferentially drive *Foxa2* expression. The fact that the alternative transcript, whose expression is driven by the second promoter, is not detected in the liver of the $Foxa2^{iCre/iCre}$ sample is likely due to detection levels. With the proof of the *Foxa2* ORF being transcribed and the fact that fusions of the *iCre* transcript to the second exon cannot be detected, this is the most plausible explanation.

These results show that a hypomorph with differential expression was generated. Taking in account that the $Foxa2^{iCre}$ allele is highly expressed in liver and pancreas (Uetzmann et al., 2008), that Foxa2 was already known to play an important role in insulin metabolism in the liver (Wolfrum et al., 2004) and that deletion of one *Foxa2* allele is already sufficient to interfere with the fat metabolism, blood and plasma analyses were undertaken in cooperation with Dr. Susanne Neschen (Institute of Developmental Genetics, Helmholtz Zentrum München).

Discussion

The analysis of blood and plasma samples taken from $Foxa2^{iCre/iCre}$ and $Foxa2^{iCre/+}$ mice compared to $Foxa2^{+/+}$ mice revealed evidence that the generated hypomorphic allele behaves like the Foxa2 knock-out allele in regard to liver metabolism (Wolfrum et al., 2008). Homozygous $Foxa2^{iCre}$ mice starved over night showed a significantly decreased high-density lipoprotein (HDL) level. It was shown that deletion of a single Foxa2 allele leads to decrease of Foxa2 expression in the liver and as a consequence to a lower HDL blood level under starving conditions (Wolfrum et al., 2008). The HDL level in the plasma is directly correlated to the Foxa2 expression level in the liver (Wolfrum et al., 2008). Foxa2 activation in the liver mediated via an insulin-dependent mechanism leads to increased oxidation and release of fatty acids (in form of triacylglycerols) (Wolfrum et al., 2004; Wolfrum and Stoffel, 2006; Wolfrum et al., 2008). Insulin in the blood plasma causes phosphorylation of Foxa2 protein which causes exclusion from the nucleus and thereby inhibits Foxa2 activity (Wolfrum et al., 2004; Wederell et al., 2008). A low level of Foxa2 protein in the liver of $Foxa2^{iCre/iCre}$ mice under starving conditions – and consequently under low plasma insulin levels – should therefore not lead to the same expression levels of Foxa2 target genes and subsequently to relatively low levels of liver lipid metabolites compared to wild-type mice. Additionally, deletion of Foxa2 in the liver leads to alterations of the bile acid homeostasis that can result in stress of the endoplasmatic reticulum and liver injury (Bochkis et al., 2008). It could be interesting to have a closer look at $Foxa2^{iCre/-}$ mice and $Foxa2^{iCre/iCre}$ mice under a cholic acid diet as these mice should be sensitized due to an expected salt accumulation in the bile ducts (Bochkis et al., 2008). This accumulation of salt could lead to gallstones: aggregates of insoluble metabolites. Furthermore, an increased salt concentration in the periphery can lead to higher uptake of salt in the kidneys where it causes oxidative stress and increases urinary albumin excretion preferentially in obese patients (Verhave et al., 2004). $Foxa2^{iCre/iCre}$ mice could therefore also be tested for higher urinary albumin excretion.

Taking together all aspects of the recently published results – increased adiposity on a high fat diet, altered HDL and triglycerides plasma levels – it is possible that Foxa2 is a key factor in the metabolic syndrome. The metabolic syndrome involves a combination of several disorders that increases the risk for diabetes or cardiovascular disease. The prevalence for this disease is estimated at 25% in the USA and is thought to be similar in most industrial nations (Ford et al., 2002). The $Foxa2^{iCre/+}$ mouse line could help to investigate the genetic background up- and downstream of Foxa2 and the different pathways influencing metabolism.
One first step could be the comparison of expression data in wild type, $Foxa2^{iCre/+}$ and $Foxa2^{iCre/iCre}$ mice.
Although the data of the plasma analysis are only preliminary because of the low number of mice that were analysed so far the results are in line with what was expected based on the Western expression data (see results) and former findings (Wolfrum et al., 2008).

Recently it was shown that Foxa2 plays an important role in the generation of dopaminergic neurons during embryogenesis and has been associated with the progressive loss of these neurons in the *substantia nigra* of the adult organism in Parkinson's disease (Kittappa et al., 2007).
Overexpression of Foxa2 in tissue culture triggered the generation of six times as many dopamine-producing nerve cells as is normally present (Kittappa et al., 2007). In vivo, precursor cells of dopaminergic neurons are found in the floorplate of the neural tube, a cell population with organizer potential and high Foxa2 expression. Also Foxa2-iCre expression could be detected on high levels within the floorplate even before the onset of neurogenesis at E10.5 (see figure 19c-f; Robinson and Dreher, 1990).
Kittappa et al. showed that mice heterozygous for Foxa2 develop Parkinson's disease when they become 18 month old which is akin to the age at which humans are most often affected (Kittappa et al., 2007). Up until now $Foxa2^{+/-}$ mice are the best experimental model systems for Parkinson's disease reflecting the loss of dopaminergic neurons in human patients. There is also striking evidence that the generated $Foxa2^{iCre}$ mouse line develops Parkinson's disease: a 14 month old heterozygous male mouse backcrossed to C57Bl/6 for four generations occasionally showed circling behaviour and had trouble to keep balance but rather fell aside (observation). An alteration in Foxa2 protein concentration in $Foxa2^{iCre/+}$ or $Foxa2^{iCre/iCre}$ mice in dopaminergic neurons has not yet been determined. Regarding the aspect that the Cre shows high activity in the floorplate, the source of these specialized neurons, a severe decrease of Foxa2 protein in this cell population is most likely (comparing it to the expression of Foxa2 protein in the liver, an organ with high Cre activity; see figure 20b). Due to the fact that the

Discussion

Foxa2$^{iCre/+}$ mouse showed Parkinsons-like behaviour at only 14 months old and was only crossed back to C57Bl/6 for four generations one would expect a more penetrant and severe phenotypic behaviour after 18 months and a pure C57Bl/6 background.

If Parkinson's disease in this mouse line can be verified it will be a second good model system. As a hypomorphic allele it might reflect the natural situation more accurately. Furthermore, it can be used as a tool to analyze factors and pathways that are involved in the development of Parkinson's disease. One can take advantage of the Cre recombination activity of this mouse line in the floorplate that allows for conditional deletion of different genes to be tested for their influence on this disease. Additionally, the fact that Cre is only expressed under the control of the first *Foxa2* promoter and that the two promoters are regulated in a different manner might help to elucidate the genetic pathways that act upstream of *Foxa2* in Parkinson's disease.

For the reasons listed here the *Foxa2$^{iCreΔneo/+}$* mouse line would not only be a useful tool to study embryonic development but also the impact of Foxa2 on metabolism and Parkinson's disease.

5.5. Conditional deletion of β-catenin in the domain of Foxa2-iCre recombination activity

Ubiquitous deletion of a gene can lead to embryonic lethality. To circumvent this problem and to be able to analyze its function in distinct tissues or cell populations genes can be deleted conditionally. Conditional alleles in which the open reading frame is flanked by *loxP* sites in the same orientation that can be recombined by Cre are the basis for these approaches. Necessary additional tools are recombinases, either temporally restricted (inducible) and/or tissue-specific. Crossing Cre mouse lines to conditional knock-out lines allow for the deletion of the conditional alleles only within the Cre expression domain. This way the function of the conditional genes can be analyzed in a distinct population.

To understand how the endoderm forms including the questions of how endoderm progenitors become lineage-specified, how they give rise to different cell types and organs and what signals are involved in these lineage decisions it is necessary to analyze the knock-out of key factors of the different pathways in specific cell populations (see figure 12).

It is known that Nodal/TGFβ as well as Wnt/β-catenin signalling is necessary for the induction of mesoderm and endoderm; knock-out models exist for both pathways that fail to gastrulate and fail to form mesoderm and endoderm (Conlon et al., 1994; Zhou et al., 1993; Haegel et al., 1995; Liu et al., 1999; Huelsken et al., 2000). In *Xenopus laevis* overexpression of Wnt1 leads to axis duplication and it is also true for exaggerated Wnt signalling in zebrafish (McMahon and Moon, 1989; He et al., 1997). To pattern mesoderm versus endoderm a gradient of Nodal signaling is sufficient with high and low levels, respectively (Zhou et al., 1993; Conlon et al., 1994; Alexander and Stainier, 1999; Tremblay et al., 2000; Lowe et al., 2001; Vincent et al., 2003). The Wnt/β-catenin pathway plays a role in the separation of endoderm and mesoderm (Lickert et al., 2002). β-catenin mutants, as knock-out models for the canonical Wnt signalling cascade, in the domain of Cytokeratin 19 Cre (*K19-Cre*) expression exhibit an accumulation of cardiac mesoderm at the expense of the generation of endoderm (Lickert et al., 2002). This pathway and its function during endoderm development seems to be highly evolutionary conserved, as canonical Wnt signalling is known to be necessary for endoderm formation in *Caenorhabditis elegans*. In *C. elegans* an EMS cell, the founder cell of mesoderm and endoderm, undergoes an unequal division and forms an MS cell, which in turn gives rise to mesoderm and an E cell that is the progenitor cell of the complete gut lineage. Balanced Wnt signalling is necessary for this step of differentiation: while loss of Wnt signalling leads to an equal division of the EMS cell giving rise to two MS-like daughter cells that will from mesoderm but no gut, an overexpression of Wnt leads to the production of two daughter cells with an E cell fate that exclusively form gut (Lin et al., 1995; for review see Han et al., 1997).

To specifically address the question of what influence the canonical Wnt pathway has on any differentiation processes in *Foxa2* positive progenitor cells in the early embryo, the Wnt downstream effector β-catenin was eliminated in the *Foxa2^{iCre}* expression domain by crossing *Foxa2$^{iCre/+}$* mice to *β-catenin$^{floxdel/+}$* and *β-catenin$^{flox/flox}$* mice that additionally carry a ROSA lacZ reporter gene (*β-catenin$^{flox/flox}$*; *R26$^{R/R}$*; Brault et al., 2001; Soriano, 1999; see figure 8; see figure 28). The reporter allele thereby allows for the sensitive detection of cells, cell populations

Discussion

and tissues with Cre recombination activity, most likely overlapping with the conditional deletion of *β-catenin* (see figure 12b and 12c).
Thus, deletion should occur in mesendodermal progenitor cells giving rise to mesoderm and endoderm.
As expected, deletion of just one allele of *β-catenin* (*Foxa2$^{iCre/+}$*; *β-catenin$^{flox/+}$* or *Foxa2$^{iCre/+}$*; *β-catenin$^{floxdel/+}$*) resulted in viable mice that had neither an obvious phenotype nor any difference in β-galactosidase expression compared to wild type mice (*Foxa2$^{iCre/+}$*; *R26$^{R/-}$*).
Following a model in which the deletion of *β-catenin* in the early embryo leads to an accumulation of mesoderm to the disadvantage of endoderm, the aim was to completely delete *β-catenin* in the domain of Foxa2-iCre recombination activity (Lickert et al., 2002).
Crossing *Foxa2$^{iCre/+}$*; *β-catenin$^{flox/+}$* or *Foxa2$^{iCre/+}$*; *β-catenin$^{floxdel/+}$* mice to conditional *β-catenin$^{flox/flox}$* mice did not result in embryonic lethality, as first expected, or in the case *Foxa2$^{iCre/+}$*; *β-catenin$^{flox/flox}$* of adult lethality up to approximately 6 months.
Taking into account that the *iCre* allele is relatively weak at early stages (E6.5-E8.5), possibly due to a discrepancy between Cre expression and recombination or to the chimaeric expression of the *iCre* under the control of only one of the two *Foxa2* promoters – one could argue that *β-catenin* is simply deleted too late and too inefficiently. This inefficiency could lead to chimaeric deletion and followed by selection for wild-type *β-catenin* or heterozygous cells. Either wild-type cells are responsible for the formation of tissues where canonical Wnt signalling is necessary or *β-catenin$^{floxdel/floxdel}$* cells undergo apoptosis as in the case of *β-catenin* deletion in the *Foxa3* expression domain (unpublished observation, Grapin-Botton A.).
Another aspect that supports this consideration is the fact that at least three alleles are there that would have to be recombined: two *β-catenin^{flox}* alleles and one *R26^R* allele. Of course, the efficiency of full recombination decreases with a rising number of alleles to be recombined, especially in situations with suboptimal levels of Cre expression driven by one of two possible promoters (see figure 17).
To find out if recombination takes place and/or if the cells undergo apoptosis immunostaining against β-catenin and a *tunel* assay would be the first experiments that one would have to accomplish.
As no embryonic phenotype could be found and adult mice do not show a metabolic phenotype as investigated up until now, phenotypes concerning stem cells that reside in the in the gut could be analyzed.
It has been shown for the crypt cells in the gut that canonical Wnt signalling has an impact on keeping stem cells in an undifferentiated state (Korinek et al., 1998; for review see Duncan et al., 2005), mutational activation initiates colon carcinoma (Korinek et al., 1997; Morini et al., 1997). Paneth cell maturation and the activation of their expression profile in the crypts depend on Wnt signalling (van Es et al., 2005). One can conclude from the data achieved with the ROSA reporter mice that the *iCre* in *Foxa2$^{iCre/+}$* mice is highly expressed in the progenitors of these stem cells. Chimeric loss of *β-catenin* might lead to a decrease in the number of cells in the stem cell population. This decrease in stem cells might not have an early phenotype in the gut especially if the deletion of *β-catenin* is chimeric. Due to a lower number of stem cells these mice might not be as sensitive to colon cancer (Korinek et al., 1997; Morini et al., 1997).
Since Wnt/β-catenin signalling has been implicated in self-renewal and maintenance of stem cell in various tissues (skin, blood and gut: for review see Duncan et al., 2005 and Beachy et al., 2004) it is likely that stem cells that reside in the liver might be effected. Here a loss of *β-catenin* could probably only lead to observable abnormalities in recovery from liver hepatectomy assays as there is not continuous substitution of cell like in the epithelium of the gut.
Focussing on defects that could possibly arise in metabolism when Wnt signalling is inhibited *Foxa2$^{iCreΔneo/+}$*; *β-catenin$^{flox/flox}$* mice were examined regarding their blood plasma composition to draw conclusions about possible metabolic diseases. Initial experiments show no significant alteration in the metabolism of *Foxa2$^{iCreΔneo/+}$*; *β-catenin$^{flox/flox}$* mice. One has to admit that final conclusions cannot be drawn due to the fact that the number of mice was too low (one mouse for the knock-out model and one for the heterozygous state and wild type).
To achieve an early deletion by reducing the number of alleles to be recombined *Foxa2$^{iCreΔneo/+}$* mice were crossed to *β-catenin$^{floxdel/+}$* mice. The offspring were then mated to *β-catenin$^{flox/flox}$*; *R26$^{R/R}$* mice in order to have *Foxa2$^{iCreΔneo/+}$*; *β-catenin$^{floxdel/flox}$*; *R26$^{R/+}$* mice for further analysis. These mice then have the advantage in that they only carry two

Discussion

alleles for recombination, allowing for more efficient and fast recombination at early stages.
Also these mice are viable until at least day P5. They are not found in a Mendelian ratio, but this can be due to the *β-catenin^{floxdel}* allele alone, as it also does not seem to follow the Mendelian rule (see chart 3). The organs of the mice look normal and there is also no difference in the staining for β-galactosidase. To further analyze these results a staining for *β-catenin* in combination with β-galactosidase would be the first step to prove complete deletion of *β-catenin* in those organs that are expressing or that have expressed *iCre*.
A stronger Cre line under the transcriptional control of the *Foxa2* promoter, like the newly established Foxa2-2A-iCre mouse line (unpublished observations; Moritz Gegg and Dr. Ingo Burtscher; Institute of Stem Cell Research, Helmholtz Zentrum München), and its regulators would be interesting to compare the results to those achieved with the *Foxa2$^{iCreΔneo/+}$* line. Possibly a stronger phenotype might be observed, if the recombination activity of the Cre starts earlier and is less chimeric. This way the phenotype occurring with the deletion of *β-catenin* in the *K19-Cre* expression domain might be recapitulated more accurately (Lickert et al., 2002).
However, it is possible that the chimeric and late deletion of *β-catenin* by using the *Foxa2$^{iCre/+}$* mouse line allows for analyzing later phenotypes due to deletion of *β-catenin* in other tissues. E.g., experiments in zebrafish show that inhibition of the Wnt/β-catenin pathway promotes the ventral midline to differentiate with hypothalamic identity rather than floorplate (Kapsimali et al., 2004). Ventral midline cells in the neural tube usually form floorplate throughout most of the central nervous system. In the anterior forebrain they differentiate with hypothalamic cell identity in the normal situation. Expression and recombination activity of Foxa2-iCre in the floorplate and deletion of *β-catenin* in these cells could also lead to an increase of hypothalamic fate of floorplate cells in mice. This might subsequently have influence on the number of cells in those cell types arising from the floorplate, e.g. dopaminergic neurons. A decrease in the number of these specialized cells could lead to Parkinson's disease later during adulthood.
More investigation will be necessary to find out if the deletion of *β-catenin* is sufficient and has a phenotype in adult mice.

5.6. Establishment and characterization of an *in vitro* ES cell assay – steps towards an *in vitro* system for embryonic development

The basic questions in stem cell research are how stem cells keep the ability to self-renew, how they are maintained into adulthood and what signals are necessary to accomplish the generation of specific cell types in *in vitro* differentiations of pluripotent (embryonic) stem cells. It is necessary to have model systems to be able to address these questions.
Especially *in vitro* differentiations have gained importance during the last decade in regard to cell replacement therapies. One reason for this obviously is that they offer the chance to produce a large amount of cells could solve the problem of the discrepancy of supply and demand of donor organs one day. Since pluripotency was successfully induced in terminally differentiated cells (induced pluripotent stem cells = iPS cells) showing that it is – in principle – possible to use a patient's own cells for replacement therapies by inducing pluripotency and subsequently differentiating these pluripotent cells into any cell type that is needed (Takahashi et al., 2007; Takahashi and Yamanaka, 2006; Nakagawa et al., 2008).
To understand the differentiation of endoderm cells and to be able to analyze the impact of different factors on endoderm development in a model system, an *in vitro* differentiation system for endoderm was successfully established. The basis of this differentiation system is the adherent co-culture with Wnt3a-expressing feeder cells in a serum-free medium supplemented with human activin A.
As previously published from other groups (Kroon et al., 2008; D'Amour et al., 2006) the adherent differentiation of ES cells into endoderm under serum-free conditions using activin A, a Nodal/TGFβ-signal (Tada et al., 2005; Yasunaga et al., 2005), can be optimized by supplementing the growth medium with Wnt3a, an activator of the canonical Wnt-signalling, as shown by comparative results of the differentiation on Wnt3a and β-galactosidase expressing feeder cells (see figure 35b). In fact, this member of the Wnt family as well as Nodal are expressed in the early embryo at the posterior side, where the endoderm arises during gastrulation (Conlon et al., 1994; Liu et

al., 1999). This correlation between *in vivo* and *in vitro* data is an example of the strong connection between the two differentiation processes. It clearly displays that the knowledge gained in one of the systems can be directly transferred to the other and vice versa. The close relation between *in vivo* and *in vitro* differentiation has already been shown for pancreatic endoderm and holds true for hepatocytes as well (Kroon et al., 2008; D'Amour et al., 2006; Agarwal et al., 2008).

There are other factors important for an efficient differentiation into the endodermal lineage that have yet to be identified. The best indication for this is that both the differentiation on Wnt3a overexpressing NIH3T3 cells as a coculture system as on β-galactosidase expressing NIH3T3 cells is more efficient when compared to differentiation on collagen IV coated dishes. These data are supported by findings in *in vitro* differentiations of human ES cells into endoderm; also in this case an increase in the efficiency of endoderm differentiation using feeder co-cultures could be observed (Zhou et al., 2008). The coculture conditions might mimic the *in vivo* situation better than collagen IV which is only one component of the extracellular matrix.

Using Wnt3a overexpressing NIH3T3 cells for differentiation inherits the consequence that the Wnt signal always stays the same. *In vivo*, only the posterior endoderm is exposed to continuously high Wnt3a signals, while the anterior endoderm (hepatic/ pancreatic endoderm), due to its movements rather has contact with Wnt inhibitors (Dessimoz et al., 2005; McLin et al., 2007). In line with these thoughts, the semi-quantitative RT-PCR showed upregulation of posterior markers (e.g. *IFAB-P* and *Cerl*; see figure 34) and downregulation of anterior markers (e.g. *Pdx1* and *Nkx2.1*, see figure 34). Although real-time PCR would be the method of choice to validate these preliminary results, they do serve as evidence for the differentiation of posterior endoderm. Also the fact that differentiating cells first upregulate *T*, *Foxa2* and *Sox17*, but subsequently downregulate *T* again while the expression of *Sox17* and *Foxa2* stays at the same level (see chart 4), indicates that the established system is comparable to the differentiation of posterior structures *in vivo* (Takaca et al., 1994; Kanai-Azuma et al., 2002; Ang and Rossant, 1994; Weinstein et al., 1994). Even the spatial resolution of the expression start of *Foxa2* versus *Sox17* at a single cell level using live imaging techniques supports this model (see figure 36) and can also be used as an argument against the theory that it might be visceral endoderm differentiated with this method. *Sox17* expression is turned on after *Foxa2* has already been expressed for 10h (see figure 36), a similar time delay of expression that is also true for the definitive endoderm *in vivo*. Observations made on long-term cultures showed that the cells *in vitro* even seem to build up tube- (gut-) like structures (figure 35e-g') that have already been observed by other groups (Torihashi et al., 2006; Matsuura et al., 2006; Yamada et al., 2002).

Summarizing the results, the differentiation system is a useful tool for testing different aspects of development *in vitro*. New cell lines can easily be tested before undergoing laborious effort to produce a mouse line. Additionally, chemicals and signalling proteins can easily be tested *in vitro* on a large scale (e.g. 384-well) regarding their potential to inhibit or sustain the differentiation of ES cells into *Foxa2* and *Sox17* positive cells.

5.7. Fusion proteins as miRNA-sensors – a tool to test micro RNAs *in vitro*?

MiRNAs are known to play important roles during embryonic development (for review see Cheng et al., 2005). The current model of miRNAs action is the inhibition of translation of the target mRNA by binding to its 3'-UTR (see figure 11 and 37; Lee et al., 1993; Olsen and Ambros, 1999; Dugas and Bartel, 2004). There is also evidence that miRNAs destabilize their target mRNAs by binding to the 3'-UTR and that some of the miRNAs are able to influence mRNA levels by using the siRNA pathway. They can directly bind to the ORF of the mRNA and lead to its degradation (see figure 11; Makeyev et al., 2007).

It is important to identify the target genes of single miRNAs to unravel their function and role during development or in the adult organism and to understand how miRNAs regulate different pathways. Today, there are many predictive software tools for miRNA target mRNAs available. However, these computational prediction algorithms allow for the estimation of up to 300 target mRNAs per miRNA (Brennecke et al., 2005; Bartel, 2004). This number impressively demonstrates why experimental approaches will be necessary to identify and verify single miRNAs and their targets.

One approach to test the miRNAs and their predicted target could be the use of knock-ins of fluorescent fusion

Discussion

proteins to certain genes that should be tested as possible targets. Fluorescent fusion proteins in general still have the same ORF as their non-targeted counterpart with, of course, additional nucleotides coding for the fluorescent protein. Another shared feature of both the fusion and native proteins is the 3'-UTR. In principle, fusion proteins carry exactly the same target sequences for miRNAs as the original (not targeted) mRNA. Theoretically, miRNAs that target a special mRNA should also target the fluorescent fusion-mRNA and therefore lead to its degradation or blocking of its translation into protein (see figure 37). No matter how the miRNA acts, the concentration of protein produced should decrease for both mRNAs. In the case of the fluorescent fusion a decrease of the fluorescence should be directly observable using fluorescent microscopy unless the fusion protein has a different secondary structure that could lead to masking of the miRNA target sites and therefore leave the mRNA translation unaffected.

In the work presented here, the impact of different miRNAs on the endoderm *in vitro* differentiation was tested. Therefore stable miRNA ES cell clones were differentiated *in vitro* according to the established protocol and analyzed for protein expression.

Using different software and tools, Dominik Lutter (Institute for Bioinformatics) predicted that miR335 targets *Sox17* and *Foxa2* mRNA. Additionally he found out that the genomic location of miR335 overlaps with the gene *Mest* (*mesoderm specific transcript*), a gene expressed predominantly in the mesoderm (Kaneko-Ishino et al., 1995, Lefebvre et al., 1998). In the cardiac mesoderm before looping, *Mest*, also known as *Peg1* (*paternally expressed gene 1*), is expressed uniformly throughout the heart tube. While the looping process proceeds expression becomes restricted to the ventricular myocardium and can also be detected in the endocard of the ventricles (King et al., 2002). At E10 the expression of *Mest* starts in the myocardium of the future right appendage and is expressed in both the left and the right at E11, but is still stronger in the right (King et al., 2002). Endocardial cells continue to show expression throughout the heart (King et al., 2002). *Mest* might have a function in the compaction of the left ventricle (King et al., 2002). Another, more recent study shows that *Mest* has an impact on fat mass expansion and suggests that *Mest* is an enzymatic component of a mechanism in the endoplasmatic reticulum that facilitates the uptake of fat into adipocytes for storage (Nikonova et al., 2008). This function of *Mest* could also explain why *Mest* knock-out mice that are viable suffer from growth retardation (Lefebvre et al., 1998).

An evaluation of expression levels of genes and the expression of the corresponding intronic miRNA could identify a significant correlation between the expression of the gene *Mest* and its corresponding miRNA, miR335 in multiple myeloma (Ronchetti et al., 2008). Predictions and transcriptional profiles associate *Mest*/miR335 with plasma cell homing and/or interaction with the bone marrow microenviroment in this study. In human, miRNA335 along with miR126 were only recently identified as suppressors of metastasis in breast cancer (Tavazoie et al., 2008). MiR335 was shown to target Sox4, a hematopoietic progenitor cell transcription factor and an extracellular matrix component in this aspect thereby suppressing metastasis and migration (Tavazoie et al., 2008).

This coincidence of reported *Mest* expression in the mesoderm (Lefebvre et al., 1998; King et al., 2002) and miR335 targeting predictions for *Foxa2* points to a *Mest*-correlated expression of miR335 in the mesoderm (compared to multiple myeloma; Ronchetti et al., 2008) where it targets and subsequently negatively regulates activators of the endodermal pathway (*Foxa2*) and thereby suppresses endoderm differentiation in the mesoderm.

To test the miRNA target prediction an expression vector was used that allowed for the expression of miRNA and a reporter protein at the same time. This vector was generated by Chung et al. (2006) as a vector for efficient overexpression and processing of miRNAs. Driven by a ubiquitous polymerase type II promoter the coupled expression of a miRNA and a protein with the same transcript is possible. Additionally, the promoter would allow for the expression of multiple miRNAs compared to commonly used polymerase III promoter-driven shRNA expression vectors that are only able to express a single miRNA from one promoter (Yu et al., 2003; Jazag et al., 2005). MiRNAs and proteins, even if their sequence is transferred with the same vector, might not have the identical expression level if their expression is driven by different promoters. In the pUI4-vector generated by Chung et al. (2006) the miRNA is located within an intron of an artifical gene (sequence with exon-intron structure), imbedded in a sequence called SIBR (*synthetic inhibitory BIC-derived RNA*) cassette. This cassette is part of the *BIC* gene, a non-protein coding gene that carries an evolutionary conserved non-coding RNA involved

Discussion

in lymphoma and other types of cancer (for detailed information see Introduction, 2.5.1.; Clurman et al., 1989; Eis et al., 2005; van den Berg et al., 2003; Kluiver et al., 2005; Iorio et al., 2005; Yanaihara et al., 2006). The SIBR cassette was shown to be the minimal sequence for efficient production of miRNAs (Chung et al., 2006). For overexpression miRNAs are cloned into the SIBR cassette by exchanging their loop with the loop of miR155, known to be processed efficiently in the miRNA pathway. The last exon of the artificial gene of the pUI4 vector comprises the open reading frame of either GFP (green fluorescent protein) or puromycin N-acetyl-transferase (which confers puromycin resistance). Therefore, it can either indicate the level of expression of the miRNA, because it is directly related to the level of fluorescent protein expression in the first case, or it can be used to obtain stable clones using the vector confers puromycin resistance.

In order to have both, the indication for the miRNA expression level on a fluorescent basis and a resistance marker to be able to select for stable integrations, the vector was altered by exchanging the ORF of GFP with the ORF of H2B-CFP-IRES-Puro (see figure 38a).

For functional analysis miR335 was subcloned into the altered expression vector and stably overexpressed in Sox17-Cherry fusion ES cells. Along with miR335, eight other miRNAs were subcloned into the same vector: miR26a, miR192, miR194-1, miR215, miR335 Sox17-optimized, miR335 Foxa2-optimized, miR350, and miR465a. MiR26a was predicted to target Sox17, miR350 and miR465a to target Foxa2 using the same tools used for the miR335 target prediction. miR192, miR194 and miR215 are miRNAs expressed in the endoderm. Logically, miRNAs that are expressed in certain tissues should support their specific lineage differentiation and suppress differentiation markers and activators of other tissues, therefore these miRNAs should not interfere with Sox17 and Foxa2 expression.

Interestingly, the Foxa2 optimized miRNA335 did not alter the expression of Foxa2 protein level while the natural miR335 did. In two independent clones miR335 overexpression resulted in a decrease of fluorescence after immunostaining for Foxa2. With the help of statistical tools (like the translation of pictures into a binary form) the interpretation of the results is more objective compared to cell counting. In this way, many pictures with information about fluorescence A can be analysed as a whole and compared to the same number of pictures of fluorescence B in a shorter period of time. Although the final proof of the correct processing of miR335 is missing, the results of the comparison of binary pictures of Foxa2 staining and miRNA/H2B-CFP expression already show an influence on Foxa2 expression and differentiation. It will be interesting to see what these cells are able to give rise to in the embryo in completely ES cell derived embryos by tetraploid complementation. The negative regulation of Foxa2 expression would only have an impact on the embryonic portion but would not influence the extraembryonic regions. This way it would be possible to study the effect on the tissues that give rise to the embryo proper. The decrease in Foxa2 protein level could possibly lead to a phenotype in the endoderm, in one of the organizer tissues (node or notochord) or both (see also $Foxa2^{-/-}$). Since miRNA overexpression, equitable with a knock-down, does most likely not reflect the complete knock-out another possibility is that only minor effects and defects would be found later in development or adulthood, analogous to the $Foxa2^{+/-}$ or $Foxa2^{iCre/iCre}$ phenotype (Wolfrum et al., 2004, see results and discussion of the hypomorphic $Foxa2^{iCre}$ allele). Regarding the fact that miR335, besides Foxa2 expression, might also influence the expression of different other genes that are not identified it might repress endoderm formation and support mesodermal development. Increased expression of factors forcing cells to differentiate into the mesodermal lineage could lead to a decrease in the size of the somites or as in the case of miR1 overexpression to decreased mass of cardiac tissue (Chen et al., 2006; Zhao et al., 2005). Of course, it cannot be excluded that miR335 overexpression leads to increase of these tissues; but a decrease of Foxa2 protein in a model system in which Foxa2 is an endodermal stem cell marker should rather support differentiation than proliferation and self-renewal. Taking in account that Mest is expressed in the endocard of the developing heart (King et al., 2002) and that miR335 expression is most likely correlated with Mest expression. Expression in the endocard could be important for the suppression of Foxa2 in endocardial cells during later heart development from E9.0 on. As known from the lineage tracing performed with $Foxa2^{iCre/+}$ mice (see figure 21) and other other Cre lines (Foxa2-2A-iCre mice, Moritz Gegg, unpublished observations; $Foxa2^{mem}$ mouse line, Park et al., 2008) Foxa2 is expressed in the progenitors of the endocard, the secondary heart field, and might have to be downregulated in this cell population during further development. Regarding this aspect embryos overexpressing

Discussion

miR335 might have a level of *Foxa2* expression in endocardial progenitors that is too low to specify them. They might suffer from severe defects in the inner lining of the heart and exhibit a mesodermal (muscular) phenotype rather than an epithelial one or at least may not be capable of fulfilling all their functions properly.

It might be also interesting to see if miR335 overexpressing mice are viable and whether they exhibit less metastasis in cancer, since miR335 is shown to have an anti-metastatic and anti-migrating effect (Tavazoie et al., 2008). This anti-migrating effect might also have impact in the developing embryo and in overexpression assays, not only in the heart where certain cell populations have to migrate to allow for turning but also during gastrulation and the formation of the three germ layers and the elongation of the antero-posterior axis. Migration defects could additionally be analyzed on a cellular level in *in vitro* systems using live imaging. Comparing data from *in vitro* studies of different miRNA transgenic ES cell lines might help to give more sophisticated predictions.

Recent findings additionally lead to the conclusion that *Foxa2*, *Mest* and miR335 are most likely components of a self-regulating feedback loop (see figure 46). First, the *Mest* promoter region has several putative Foxa2 binding sites predicted by the Genomatix software tool. Second, *Mest* appears to be downregulated in *Foxa2* knock-out embryos based on an expression screen that compared *Foxa2* wild type and $Foxa2^{-/-}$ mouse embryos at E7.5 (Tamplin et al., 2008, in press). In the secondary heart field Foxa2 might be necessary to induce *Mest* while it has to be downregulated for later development in the endocard by miR335. This is supported by the lineage tracing of $Foxa2^{iCre/+}$ mice compared to *in situ* data (see figure 21; Monaghan et al., 1993; Sasaki and Hogan, 1993; Ang and Rossant, 1994). *Foxa2* is expressed in the progenitors of the endocard but is absent in the heart at later stages. *Mest*, due to its enzymatic function in fat metabolism, could be important for the enormous growth of the heart at these early stages, while the function of miRNA335 is predominantly to downregulate *Foxa2* to allow for the further differentiation of the cells in the endocard.

In light of the enzymatic function of *Mest* and its involvement in fat metabolism, as well as the finding that *Mest* knock-out mice show growth retardation (Lefebvre et al., 1998) and the fact that $Foxa2^{+/-}$ mice exhibit fat metabolism defects and that *Foxa2*, directly or indirectly, activates *Mest* (Genomatix predictions and Tamplin et al., 2008, in press) it would be interesting to investigate (using RT-PCR, *in situ* analysis and immunostainings) if *Mest* and miR335 are expressed in the liver along with *Foxa2* at the same time. Due to the fact that *Foxa2* regulates fat metabolism and that *Mest* is also involved, the same regulative mechanisms (activation of *Mest* by *Foxa2* and subsequent downregulation of *Foxa2* by miR335) might be conserved in the liver. Conditional deletion of one *Mest* allele in $Foxa2^{iCre}$ mice could lead to a compound phenotype similar to that of the $Mest^{-/-}$ mice if *Foxa2* and *Mest* are components of the same pathway.

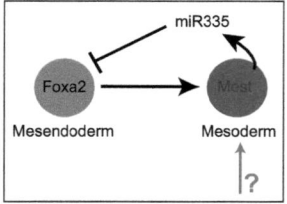

Figure 46: Model of a feedback loop of Foxa2 and miR335
Figure 46 illustrates how Foxa2 first activates Mest and thereby miR335 and is then consequently inhibited by miR335. Only in certain cell types where *Foxa2* and *Mest* are both expressed this feedback loop can serve as a model. In other cell types Mest expression might be activated by other transcription factors, possibly T or Fgf (indicated by grey arrow).

It is also interesting to note that the level of *Sox17* expression (although it was also a predicted target) did not seem to be effected. This can be due to sensitivity levels of the fluorescence intensity and missing resolution at the Axiovert microscope compared to a confocal microscope. Another possible explanation is that the prediction is only partially true. One indication for this alternative is that another combination of bioinformatic tools now lead to the exclusion of *Sox17* from the predicted target gene pool of miR335 (software tool "Targetspy", Martin Sturm, unpublished finding).

None of the other miRNAs predicted to target *Foxa2* or *Sox17* seemed to negatively regulate *Foxa2* or *Sox17* expression. Of course, one explanation for the missing effect could be that the miRNAs have simply no influence on the mRNA translation or degradation of these two genes. As mentioned before, it can also be due to sensitivity levels that cannot be resolved with the technique used (fluorescence microscopy). The miRNAs might simply not be processed correctly within the miRNA pathway or target sites of the mRNA that could be masked in the fluorescent fusion variant of the mRNA. Only Northern blot and real-time PCR analysis of *Foxa2* mRNA concentrations could give an answer to that question, in the case of Sox17 immunostainings against the fluorescent fusion protein of the wild type protein might help to increase the fluorescent signal and allow for a conclusion.

Foxa2 expression seemed to be increased by miR194 overexpression. Since miR194 is expressed in the endoderm (Landgraf et al., 2007) and should therefore support its formation, the prediction was that it either positively influences *Foxa2* expression or has no effect at all. Compared to other miRNAs that are also expressed in the endoderm, miR215 and miR192, *Foxa2-Venus* expression was upregulated. There are different possibilities for this phenomenon. Clonal variation can always be used as an argument in *in vitro* assays, especially if only one clone is tested. For miR194-overexpression two independent clones (see Results) show the same effect. In this case the proof of the miRNA processing and activity cannot be measured by the regulation of the mRNA levels of its targets, because the influence on *Foxa2* is only indirect and the real target mRNAs are not known. Using the bioinformatic tools for prediction could give some hints as to what mRNAs are necessary to focus on. The targeting of these mRNAs could be tested *in vitro* by expressing fluorescent fusion proteins and miRNAs in HEK293 cells in parallel. FACS analysis or Northern blot would be assays for the examination. Additionally, in tetraploid aggregation assays more information about the impact on *in vivo* differentiation and development could be gathered using the stable ES cell clones generated in this thesis. Once a phenotype is found, it might be possible to identify pathways that are involved and mRNAs that are eventually targeted by miR194.

Regardless what targets miR194 might have, since it indirectly upregulates *Foxa2* it is possible that its overexpression will have an opposing effect to miR335 *in vivo* using the tetraploid aggregation assay. Formation of endoderm to a greater extent due to a support of endoderm to the disadvantage of the mesodermal lineage or overgrowth in certain subpopulations of the endoderm could be possible effects. Especially in the anterior part of the embryo, where the liver differentiates under the influence of the cardiac mesoderm (Deutsch et al., 2001; Gualdi et al., 1996) and where *Foxa2* knock-out mutants show a severe defect in endoderm formation (Ang and Rossant, 1994; Weinstein et al., 1994) one might expect to see clear differences under miR194 overexpression. E. g., if it represses (cardiac) mesoderm formation it might support the formation of the liver and other more anterior endodermal structures indirectly by repressing Fgf signaling from the adjacent mesoderm (Dessimoz et al., 2006); and if it supports the expression of genes that force endoderm proliferation and differentiation it might also support e. g. liver or pancreas formation directly (Zaret, 2008).

In the next step the miRNA constructs should be introduced into double targeted *Foxa2*-Venus fusion and *Sox17*-Cherry fusion (and vice versa: *Foxa2*-Cherry fusion and *Sox17*-Venus fusion) knock-in ES cells because these will allow for an even easier and faster read-out *in vitro* and *in vivo*. Foxa2 and Sox17 protein decrease can be visualized directly and immediately correlated to miRNA expression without further immunostainings *in vivo* using live-imaging techniques. As published recently, supplementing the differentiation medium with a drug called enoxacin could enhance the knock-down effect by promoting the biogenesis of mRNAs (Shan et al., 2008). An even stronger effect of miR335 or miR194 expression might be detectable but should be controlled tightly because it is not known what other endogenous miRNAs might be upregulated using enoxacin.

The results show that the *in vitro* differentiation system can be used as a test system for miRNAs and that it is a fast tool to analyze miRNAs *in vitro* in an environment that mimics the *in vivo* situation as closely as possible. They further clear the way for subsequent tests *in vivo* in tetraploid aggregation assays.

6. Material and Methods

6.1. Material

6.1.1. Equipment

Centifuges	5417 R (Eppendorf AG, Hamburg)
	5417 C (Eppendorf AG, Hamburg)
	5804 R (Eppendorf AG, Hamburg)
	Haereus Rotanta 460R (Thermo Fisher Scientific Inc., Waltham)
	Hettich Universal 30F (Andreas Hettich GmbH & Co. KG, Tuttlingen)
	1-14 (Sigma Laborzentrifugen GmbH, Osterode am Harz)
	Galaxy Mini (VWR International GmbH, Darmstadt)
Incubation systems/ovens	Shaking incubator; 37°C bacteria (Shel Lab, Sheldon Manufacturing, Cornelius)
	TH-30 and SM-30; 32°C bacteria (Edmund Bühler GmbH, Hechingen)
	65°C Southern Blot
	Thermomixer comfort (Eppendorf AG, Hamburg)
	Shake'n'Stack (ThermoHybaid, Thermo Fisher Scientific Inc., Waltham)
Electroporation system	BioRad Gene Pulser Xcell (BioRad Laboratories, München)
Power suppliers	Power Pack Basic (BioRad Laboratories, München)
Agarose gel chamber	Midi 450 (Harnischmacher, Kassel)
Gel documentation system	UV-Transilluminator (Biorad, München)
	Gene Flash (Syngene Bio Imaging, Synoptics Ltd, Cambridge)
Pipettes	1000µl/ 100µl/ 20µl/ 10µl Eppendorf Research (Eppendorf AG, Hamburg)
	Pipettboy accu-jet and accu-jet® pro (Brand GmbH & Co. KG, Wertheim)
Photometer	BioPhotometer (Eppendorf)
	ND-1000 Spectrophotometer (NanoDrop, (Thermo Fisher Scientific Inc., Waltham)
Polyacrylamid gel preparation	(BioRad)
Polyacrylamid gel chamber	Mini Trans-Blot® Cell (BioRad GmbH, Heidelberg)
Western Blot semi-dry system	Trans-Blot® SD, Semi-Dry Transfer cell (Biorad, Heidelberg)
PCR machines	Px2 ThermoHybaid (Thermo Fisher Scientific Inc., Waltham)
	PXE0.2 Thermo Cycler (Thermo Fisher Scientific Inc.Waltham)
FACS	FACS Calibur (Bector and Dickinson and Company, Franklin Lakes)
Balances	Scout™ Pro (OHAUS)
	(Sartorius)
Vortexer	Vortexer (VWR international GmbH, Darmstadt)
Rotator/tumbler	VSR 23 (Grant BOEKEL, VWR international GmbH, Darmstadt)
	Roller Mixer SRT1 (Bibby Scientific (Stuart), Staffordshire, GB)
Water bath	VWR

Material and Methods

pH meter	pH211 Microprocessor pH Meter (HANNA instruments Deutschland GmbH, Kehl am Rhein)
Pumps	LABOPORT ® (neoLab Migge Laborbedarf-Vertriebs GmbH, Heidelberg)
Hybridisation tubes	Hybridizer HB 100 (ThermoHybaid, (Thermo Fisher Scientific Inc., Waltham)
Film cassettes	Hypercassette (Amersham, GE Healthcare GmbH, München)
Developing machine	AGFA Curix 60 developing machine (AGFA HealthCare GmbH, Bonn)
Radiation monitor	Berthold LB122 radiation monitor (BERTHOLD TECHNOLOGIES GmbH & Co. KG, Bad Wildbach)
Microscopes	Axiovert 200M (Carl Zeiss AG, Göttingen)
	Lumar.V12 (Carl Zeiss AG, Göttingen)
	MS5 (Leica Microsystems GmbH, Wetzlar)
	TCS SP5 (Leica Microsystems GmbH, Wetzlar)
Cameras	AxioCam MRc5 (Carl Zeiss AG, Göttingen)
	AxioCam HRm (Carl Zeiss AG, Göttingen)
Microwave	700W (Severin Elektrogeräte GmbH, Sundern)
Stirrer	STIR (VWR international GmbH, Darmstadt)
Cross-linker	UV Stratalinker 1800 (Stratagene)
Microtome	Microm HM 355 S rotation microtome (Thermo Fisher Scientific Inc., Waltham)
Glassware	Schott-Duran (Schott, Mainz)
Plastic ware	(VITLAB GmbH, Großostheim)
Counting chamber (cells)	Neubauer (LO - Laboroptik GmbH, Friedrichsdorf)
Freezer	20°C (Liebherr Hausgeräte Ochsenhausen GmbH, Ochsenhausen)
Fridge	4°C (Liebherr Hausgeräte Ochsenhausen GmbH, Ochsenhausen)

6.1.2. Consumables and kits

Consumables

50ml/ 15ml tubes	(Becton and Dickinson and Company, Franklin Lakes; Sarstedt, Nürnbrecht)
14ml tubes	BD Labware (Becton Dickinson GmbH, Heidelberg)
2ml/ 1,5ml safe-lock reaction tubes	(Eppendorf AG, Hamburg)
0,2ml tubes	(Eppendorf AG, Hamburg)
15cm/ 10cm/ 6cm dishes	nunc (Thermo Scientific Fisher, Wiesbaden)
6-well/ 12-well/ 24-well/ 48-well plates/ 96-well plates (straight/conical)	nunc (Thermo Scientific Fisher, Wiesbaden)
10cm bacterial plates	BD Falcon™ (Becton Dickinson GmbH, Heidelberg)
Embedding cassettes	(Carl Roth GmbH & Co. KG, Karlsruhe)
Embedding moulds	(Carl Roth GmbH & Co. KG, Karlsruhe)
50ml/ 25ml/ 10ml/ 5ml/ 2ml/ 1ml plastic pipettes	(Greiner bio-one, Frickenhausen)
Pasteur pipettes, plastic	transfer pipettes (Carl Roth GmbH & Co. KG, Karlsruhe)

Pasteur pipettes, glass	15cm/ 23cm (LABOR-BRAND, Gießen; Hirschmann Laborgeräte GmbH & Co. KG, Eberstadt)
Parafilm	Parafilm (Pechiney Flastic Packaging, Menasha)
PVDF membrane	Immun-Blot PVDF-Membrane (BioRad Laboratories, Hercules)
Nitrocellulose membrane	(GE Healthcare Buchler GmbH & Co. KG, München)
Blotting paper	Whatman paper (GE Healthcare Buchler GmbH & Co. KG, München)
Scalpels	surgical disposable scalpels B/Braun (Aesculap AG & Co. KG Tuttlingen)
Films	Kodak BioMax MS (Sigma-Aldrich GmbH, Hamburg), Amersham Hyperfilm ECL (GE Healthcare Buchler GmbH & Co. KG, München)

Kits

QIAquick PCR Purification Kit (Qiagen Holding, Hilden)
QIAquick Gel Extraction Kit (Qiagen Holding, Hilden)
QIAgen Maxi Kit (Qiagen Holding, Hilden)
QIAgen Mini Kit (Qiagen Holding, Hilden)
RNeasy Mini Kit (Qiagen Holding, Hilden)
Labelling Kit (Roche Holding GmbH, Applied Science, Mannheim)
Nextract (Sigma-Aldrich GmbH, Hamburg)
ECL Detection Kit (Millipore Cooperation, Billerica, MA)

6.1.3. Chemicals

(Sigma-Aldrich GmbH, Hamburg, Merck KGaA, Darmstadt, Carl Roth GmbH & Co. KG, Karlsruhe)

A	Acetic acid
	Activin A, human (R&D Systems, Minneapolis)
	Acrylamide/bisacrylamide
	Agarose (Biozym Scientific GmbH, Hess. Oldendorf)
	Ampicillin
	APS
B	BCA
	BSA
	Bradford reagent
	bromine phenol blue
C	Calcium chloride
	Chloroform, 99+%
	CI (Chloroform-Isoamylalcohol: 24:1)
D	Diethylpyrocarbonate (DEPC), approx. 97%
	Dimethylsulfoxide (DMSO), >99,9%
	Dithiothreitol (DTT)
	dNTPs (Fermentas GmbH, St. Leon-Rot)
E	EDTA
	Ethanol, 96%
	Ethidiumbromide
F	Formaldehyde

Material and Methods

F	Formamide
G	Gelatine
	Glutamine
	Glutaraldehyde
	Glycerol
	G418 (Geneticin, 50mg/ml, Gibco, Invitrogen™ Cooperation, Carlsbad, CA)
H	HEPES (200mM, Gibco, Invitrogen™ Cooperation, Carlsbad, CA)
	HEPES (powder)
	HCl
I	Isopropanol, 100%
K	Kanamycin
L	L-glutamine (200mM, Gibco, Invitrogen™ Cooperation, Carlsbad, CA)
M	Magnesium chloride
	Methanol, 100%
	MEMs non essential amino acids (100x, Gibco, Invitrogen™ Cooperation, Carlsbad, CA)
	Milk powder (Becton Dickinson GmbH, Heidelberg)
	Mitomycin C
	Mounting medium
	MOPS
	β-mercaptoethanol (50mM, Gibco, Invitrogen™ Cooperation, Carlsbad, CA)
N	Nitrogen$_{(l)}$ (Linde AG, München)
	Nuclear Fast Red
O	Oligo-dT-primer (Promega GmbH, Mannheim)
P	Paraformaldehyde
	PBS (Gibco, Invitrogen™ Cooperation, Carlsbad, CA)
	PCI (Phenol-Chloroform-Isoamylalcohol: 25:24:1; Carl Roth GmbH + Co. KG, Karlsruhe)
	Penicillin/Streptomycin (Gibco, Invitrogen™ Cooperation, Carlsbad, CA)
	Polyacrylamide
	Potassium acetate
	Puromycin
Q	Q-Solution (Qiagen Holding, Hilden)
R	RNaseZAP
	Rotihistol
S	Sodium chloride
	Sodiumdodecylsulphate (SDS)
	Sodium hydrogenic phosphate (Na_2HPO_4)
	Sodium hydroxide
T	TEMED
	TWEEN20
	Tris
	Triton X-100
	Trizol Reagent (Gibco, Invitrogen™ Cooperation, Karlsruhe)
X	X-Gal
	Xylene

Material and Methods

6.1.4. Buffers and solutions

Isolation of genomic DNA
Proteinase K lysis buffer:	100mM	Tris, pH8.0-8.5
	5mM	EDTA, pH8.0
	2%	SDS
	200mM	Sodium chloride

Plasmid preparation
P1 buffer:	50mM	Tris HCl, pH 8.0
	10mM	EDTA
	100µg/ml	RNase A
P2 buffer:	200mM	Sodium hydroxide
	1%	SDS
P3 buffer	3M	Potassium acetate, pH 5.5
QBT buffer	750mM	Sodium chloride
	50mM	MOPS, pH 7.0
	15%	Isopropanol (v/v)
	0.15%	Triton X-100 (v/v)
QC buffer:	1M	Sodium chloride
	50mM	MOPS, pH 7.0
	15%	Isopropanol (v/v)
QF buffer:	1,25M	Sodium chloride
	50mM	Tris HCl, pH 8.5
	15%	Isopropanol
TE buffer:	10mM	Tris HCl, pH 8.0
	0.1mM	EDTA
EB buffer:	10mM	Tris HCl, pH 8.0

DNA/ RNA agarose gels
TAE buffer (50x stock):	2M	Tris
	50mM	Glacial acetic acid
	50mM	EDTA
Loading buffer DNA:	100mM	EDTA
	2%	SDS
	60%	Glycerol
	0.2%	Bromine phenol blue
Loading buffer RNA (2x):	95%	Formamide
	0.025%	SDS
	0.025%	Bromine phenol blue
	0.025%	Xylene cyanol FF
	0.025%	Ethidium bromide
	0.5mM	EDTA

Southern blot
Depurination (fragments >=10kb):	1.1%	HCl in H_2O
Denaturation (all gels):	87.66g	Sodium chloride
	20.00g	NaOH
	1000ml	H_2O (final volume)

Neutralization (all gels):	87.66g	Sodium chloride
	60.50g	Tris
	1000ml	H$_2$O (final volume)
	pH7.5	with HCl conc. (approx. 11ml)
Transfer, 20x SSC (all gels):	88.23g	Tri-sodium-citrat
	175.32g	Sodium chloride
	1000ml	H$_2$O (final volume)
	pH7-8	
Hybridisation buffer:	1M	Sodium chloride
	50mM	Tris, pH7.5 (at 37°C)
	10%	Dextransulfate
	1%	SDS
	250µg/ml	Salmon Sperm DNA sonificated
	→ store aliquots à 30ml at -20°C	
Washing buffers:	a) 2x SSC / 0.5% SDS	
	b) 1x SSC / 0.5% SDS	
	c) 0.1% SSC / 0.5% SDS	
stock solutions:	a) 20x SSC	175.3g Sodium chloride
		88.2g sodium citrate
		pH7.0
	b) 20% SDS	200g SDS
		1000ml H$_2$O (final volume)

Western blot

Lysispuffer:	50mM	Tris/HCl, pH7.4
	150mM	Sodium chloride
	2mM	EDTA, pH8
	1%	Nonidet P-40
	filtrate sterile	
APS:	10%	APS (in dest. H$_2$O)
4x Tris/SDS pH8.8:	1.5M	Tris (→ pH8.8)
	0.4%	SDS
4x Tris/SDS pH6.8:	0.5M	Tris (→ pH6.8)
	0.4%	SDS
10x Tris-Glycine (Running buffer):	1.0%	SDS
	0.25M	Tris
	1.92M	Glycine
4x SDS-loading dye:	(2M DTT add freshly: 40µl to 160µl buffer)	
	200mM	Tris/HCl, pH6.8
	8%	SDS
	40%	Glycerol
	0.4%	bromine phenol blue
Buffer cathode:	25mM	Tris/HCl, pH9.4
	40mM	Glycine
	10%	Methanol
Buffer anode I:	300mM	Tris/HCl, pH10.4
	10%	Methanol
Buffer anode II:	25mM	Tris/HCl, pH10.4
	10%	Methanol
Ponceau-Lösung:	0.2%	PonceauS
	3%	TCA

10x TBST:	(washing buffer, add Tween20 freshly)	
	100mM	Tris-HCl, pH 7.4
	1.5M	Sodium chloride
	1.0%	Tween20
Blocking solution:	1:10 (v/v)	milk powder
	1g	BSA
		in 1x TBST
ECL-Lösung:	mix directly before usage	
	Solution A and B mix: 1:1	

FACS analysis

FACS buffer:	20%	FCS
		in PBS

LacZ-staining

Fixation buffer:	0.02%	NP-40
	5Mm	EGTA, pH8.0
	2mM	$MgCl_2 \times 6H_2O$
	1%	Formaldehyde
	0.2%	Glutaraldehyde
		in PBS
Washing buffer:	0.02%	NP-40
		in PBS
Staining buffer:	0.02%	NP-40
	2mM	$MgCl_2 \times 6H_2O$
	5mM	$K_3[Fe(CN)_6]$
	5mM	$K_4[Fe(CN)_6] \times 6H_2O$
	0.01%	Natriumdesoxycholat
	1mg/ml	X-Gal
		in PBS

Transfections

2x HBS buffer:	0.27M	Sodium chloride
	0.054M	HEPES
	0.001	Sodium hydrogenic phosphate (Na_2HPO_4)
	pH 7.05	

Immunostainings

TBS buffer (10x):	0.5M	Tris(hydroxymethyl)aminomethane
	1.625M	Sodium chloride
	adjust to pH 7.6 with conc. HCl	

6.1.5. Enzymes and enzyme kits

Proteinase inhibitors	(Sigma-Aldrich GmbH, Seelze)
Superscript II	(Fermentas GmbH, St. Leon-Rot)
RNase inhibitors	(Fermentas GmbH, St. Leon-Rot)
Restriction enzymes	(NEB GmbH, Frankfurt a. M.; Fermentas GmbH, St. Leon-Rot)
DNA-Polymerases	(DNA Polymerase I, Large (Klenow) Fragment, NEB GmbH, Frankfurt a. M.; M0210; *Taq* DNA Polymerase recombinant, Fermentas GmbH, St. Leon-Rot, EP0402; *Taq* DNA Polymerase; Qiagen, Hilden; 201203; *Pfu* DNA Polymerase, Stratagene, La Jolla)

Material and Methods

        RNase   (Promega GmbH, Mannheim)
        RNase-free Dnase I   (Promega GmbH, Mannheim)
        Ligase   (T4 DNA ligase; NEB GmbH, Frankfurt a. M.; M0202)
        Phosphatase   (T4 Polynucleotide Kinase, NEB GmbH, Frankfurt a. M.; Antarctic phosphatase NEB GmbH, Frankfurt a. M.)

6.1.6. Antibodies and sera

Primary antibodies
        Polyclonal goat α-*Foxa2* antibody M-20 (Santa Cruz Biotechnology, Inc., Santa Cruz)

Secondary antibodies
        rabbit α-goat IgG HRP-conjugated (Dianova GmbH, Hamburg)
        rabbit anti-goat IgG 488 (Invitrogen GmbH, Karlsruhe)

Sera
        Sheep serum (Sigma-Aldrich GmbH, Hamburg)

6.1.7. Vectors and BACs

Vectors
        pBluescript
        pL451-*loxP* (Lee et al., 2001; Liu et al., 2003; modified)
        pUI4-SIBR-GFP (Chung et al., 2006)
        pUI4-SIBR-H2B-CFP-IRES-Puro (Chung et al., 2006 ; modified)

BACs
        RPCI22-254-G2 (RZPD; 129Sv *Foxa2*-BAC; Osoegawa et al., 2000)
        RPCI22-46-A17 (RZPD; 129Sv *Sox17*-BAC; Osoegawa et al., 2000)

6.1.8. Oligonucleotides

Application	Name	Sequence
Genotyping	EP176 *Flp-e* sense	5'-CTAATGTTGTGGGAAATTGGAGC-3'
	EP177 *Flp-e* reverse	5'-CTCGAGGATAACTTGTTTATTGC-3'
	EP308 *R26R* 1	5'-AAAGTCGCTCTGAGTTGTTAT-3'
	EP309 *R26R* 2	5'-GCGAAGAGTTTGTCCTCAACC-3'
	EP310 *R26R* 3	5'-GGAGCGGGAGAAATGGATATG-3'
	EP418 *Foxa2/Sox17* genotyping 1	5'-AGCCATACCACATTTGTAGAGG-3'
	EP419 *Sox17* genotyping 2	5'-CTGCTGACCATTCTCTTGATAG-3'
	EP420 *Foxa2/Sox17* neo genotyping 1	5'-ATTGCATCGCATTGTCTGAGTAG-3'
	EP421 *Foxa2* genotyping 2	5'-CAAAACAACAAGCAGGTGACAG-3'
	EP482 *β-catenin* 1	5'-AAGGTAGAGTGATGAAAGTTGTT-3'
	EP483 *β-catenin* 2	5'-CACCATGTCCTCTGTCTATTC-3'
	EP484 *β-catenin* 3	5'-TACACTATTGAATCACAGGGACTT-3'
	EP510 *Sox17* genotyping 3	5'-CTGTGCAATTGGACTTGAATG-3'
	EP511 *Foxa2* genotyping 3	5'-GGGAGACAAGGTTTCTCTCTG-3'

Table 5: Oligonucleotides - Part I

Material and Methods

Application	Name	Sequence
Cloning	EP002 *NotI Foxa2* Exon 1A reverse	5'-NNNgcggccgcGGTCGTCAGTTACCTCAGTCCTCTTTC-3'
	EP004 *NotI Sox17* Exon 1A reverse	5'-NNNgcggccgcCGCCAGCAGTGTGAGAGGGCCATATTTCAG-3'
	EP005 *SacI-NsiI Foxa2* Exon 1A sense	5'-NNNgagctcatgcatTGTAGCCCGGATCTCTTGAAACTCACTC-3'
	EP006 *SacI Sox17* Exon 1A sense	5'-NNNccgcggGATCTGCTTGAGTGCCCACGGATCCTGTGC-3'
	EP007 *EcoRI Foxa2* Exon 1B sense	5'-NNNgaattcCAGCGGCCAGCGAGTTAAAGGTG-3'
	EP008 *HindIII Foxa2* Exon 1B reverse	5'-NNNaagcttCGGGCAGCCCATTTGAATAATCAGC-3'
	EP009 *EcoRI Sox17* Exon 1B sense	5'-NNNgaattcCACTCCTCCCAAAGTATCTATCAAGAGAATG-3'
	EP010 *HindIII Sox17* Exon 1B reverse	5'-NNNaaggttGCATTTTCTCTGTCTTCCCTGTCTTGGTTG-3'
	EP011 *SpeI* Intron-pA fwd	5'-NNNactagtAGGTAAGTGTACCCAATTCGCCCTATAG-3'
	EP012 *BamHI* Intron-pA reverse	5'-NNNggatccACGCGTTAAGATACATTGATGAGTTTGGAC-3'
	EP015 *NotI* Kozak-*iCre* sense	5'-NNNgcggccgcGCCACCATGGTGCCCAAGAAGAAGAGGAAAG-3'
	EP016 *SpeI* Kozak-*iCre* reverse	5'-NNNactagtTCAGTCCCCATCCTCGAGCAGCCTCAC-3'
	EP108 PL253-*NotI*-Oligo fwd	5'-GGCCGTTTAAACTTAATTAAGCTTGGCGCGCCATGCATTTAAAT-3'
	EP109 PL253-*NotI*-Oligo rev	5'-GGCCATTTAAATGCATGGCGCGCCAAGCTTAATTAAGTTTAAAC-3'
	Foxa2 3'-probe reverse primer	5'-CCATGGGAATGGCCTAT-3'
	Foxa2 3'-probe forward primer	5'-CTGGATATGCTCTAGAAAGGC-3'
	Foxa2 fwd 3' HR *HindIII*	5'-GGGaagcttTGCTCACACAACAACATTGCTCATGGTC-3'
	Foxa2 fwd 5' HR *NotI*	5'-GGGgcggccgcACTTCGACATGTAAGATCTCTAGACAGAAC-3'
	Foxa2 rev 3' HR *SpeI*	5'-GGGactagtCTCGAGAACCCAAGGGCCCTCAACATCAG-3'
	Foxa2 rev 5' HR *HindIII*-*NsiI*	5'-GGGaagcttcatatgATGCAGAGCACGA3TCCCTCAAAGGCAAC-3'
	loxP deletion antisense oligo	5'-gatcccgggaagttcctatactt ctagagaataggaacttctt-3'
	loxP deletion sense oligo	5'-cgaagaagttcctatctctagaaagtataggaacttcccgg-3'
	Sox17 5'-probe reverse primer	5'-NNNaccggtCGGGAATGTTTC-3'
	Sox17 5'-probe forward primer	5'-NNNaagcttTGGAAGCTAAATTAGGTTC-3'
	Sox17 fwd 3' HR *HindIII*	5'-GGGaagcttATTTGTGTGTGGTGTGCTTTACGCAGG-3'
	Sox17 fwd 5' HR *NotI-AgeI*	5'-GGGgcggccgcaccggtTACTGAGCTGGGAAG-3'
	Sox17 rev 3' HR *SpeI*	5'-GGGactagtGAATTCTAGTGCCCACCTATGCCCCCTAC-3'
	Sox17 rev 5' HR *HindIII-NsiI*	5'-GGGaagcttcatatgAGTGGGTCGGAGCTGGAGATGGAAATTC-3'
Sequencing	EP604 miR sequencing sense	5'-CTACTCTGTTGACAACCATTG-3'
	EP605 miR sequencing reverse	5'-GATGAGACAGCACAATAAC-3'
RT-PCR	EP083 *Foxa2* sense	5'-GACTGGAGCAGCTACTACGCG-3'
	EP084 *Foxa2* reverse	5'-GCTCAGACTCGGACTCAGGT-3'
	EP085 *Hex* sense	5'-AGTGGCTTCGGAGGCCCTCTGTAC-3'
	EP086 *Hex* reverse	5'-GCCCGGATCCTGACTGTCATCCAGCATTAA-3'
	EP087 *Pdx1* sense	5'-CCACCCCAGTTTACAAGCTC-3'
	EP088 *Pdx1* reverse	5'-TGTAGGCAGTACGGGTCCTC-3'
	EP093 *Sox17* sense	5'-GCGAGGTGGTGGCGAGTAG-3'
	EP094 *Sox17* reverse	5'-TCTGCCAAGGTCAACGCC-3'
	EP095 *Nkx2.1* sense	5'-CCTGGAGGAAAGCTACAAGAAA-3'
	EP096 *Nkx2.1* reverse	5'-GGTTCTGGAACCAGACTTGAC-3'
	EP142 *β-actin* sense	5'-GACGAGGCCCAGAGCAAGAG-3'
	EP143 *β-actin* reverse	5'-ATCTCCTTCTGCATCCTGTC-3'
	EP232 *Cer1* sense	5'-GGAAGAAACCTGAGACCGAAT-3'
	EP233 *Cer1* reverse	5'-AGTCCAGGGATGAAGGAACC-3'
	EP262 *IFAB-P* sense	5'-GACCGGAACGAGAACTATG-3'
	EP263 *IFAB-P* reverse	5'-CAGGCTCTGAGAAGTGAC-3'
	EP546 *Foxa2-iCre* expression 1	5'-CTTCCATCTTCACGGCTCCAGC-3'
	EP547 *Foxa2-iCre* expression 2	5'-CAGCGGCCAGCGAGTTAAAG-3'
	EP548 *Foxa2-iCre* expression 3	5'-CTTTAACTCGCTGGCCGCTG-3'
	EP567 *Foxa2-iCre* expression 4	5'-GGTCGTTTGTTGTGGCCTG-3'
	EP568 *Foxa2-iCre* expression 5	5'-GCTGGTGGCTGGACCAATGTG-3'

Table 5: Oligonucleotides - Part II

6.1.9. Probes

Foxa2 3' probe Southern blot, length: 744bp
Primer for amplification: see table 5

Sequence:
5'-CTCGAGCTGGATATGCTCTAGAAAGGCAGAAGTTTACAGTTTTTTTAATATCAGGCCTCCTTTC
TAGTCAGTGAACTTAGACTGGGTTTACCAATTTTGGTGCATGGCTCTTCCAGCTACTTGAAG
CATTGCCCCCCCTAGACCTTCCTGTGCCATTGAGACTACCTGGCTCTAGGTTGTGCCGGGAGGG
CAGCCTGTCTCAGTCTCACAGGTGTTATCCAGGTATTGGGAAACCTTGCTAGGCTAGGAACGAT
GAGCCACCTAATCTGGGGAAACATTTTAACATTGGGAATTGGGTATAATTGCATAGTTAAGGGTAAC
CCCCAAATCTTTTATTAAGAAGTTATTCTGTGGGTGGGGAGATAGGGAGGGATGGAAGGGTG
CCCTGAGCAGCTTAGCAAATGACTCCCAAAGTAGTGAAATCCCAGTGTCTCAGGAATGGTGTCTC
CCTTCTACCAGCCAGGGCAAAGCTGTTTGTTAGCTTAGGAAGCTCCTATAGGCAAACCACACTT
GAGGCCCAGGGACTGAATGGGTATTTTGTGAGCCTCCAGGAAAATACAAAGACCCCAAATAAAAC
CTCACCAATCATTTCCACCACTCTGCAGATTTTCCAAATTGACGGGTAACTGTAGAGGAGGT
CGTGTTTTGCAAAAGGAGCCTCCTCACGCTGACCTGCATCTCCTGCCCTTGAAGCTGTCCCTCC
CGCCCGCCCCCAGTCTGACTTTCCATAGGCCATTCCCATGG-3'

Sox17 5' probe Southern blot, length: 744bp
Primer for amplification: see table 5

Sequence:
5'-AAGCTTTGGAAGCTAAATTAGGTTCAGGTTAGAAATCTGTTATGCTTGAAGCGCATACAG
GAGGTAATGAGTGCTGTCAGGAGTGGCAAGTCCTATTTAACTTTGATCATTTCCATGATCACA
GAGCACATACCTTGCCCTAACCAACAATCTGGCGCCATGCAGTTAATCCTGATAGGGGATCTTGG
GAGCAGAGTTTTCTGGATGGAAAGGCACCTATTGCCTCTGCTACAAAACCAAAGCAAACTCAACG
AAACAACTCAATGTGCTTGACCCACTGTGTAGTCATCATCACTTTCACAGTCCAGGAACGGAGTAT
TACAGGATCATAGACTGAGCTGGATATTAATAAATCAGAGAAGAAGACTAAGGATGCCACACTGGG
CATCCTGGGAAGCAGTCCTACCCAGTTTGCTCTCTGGAGAGAGTGCAAGTGCTCTGCTCA
CAGTGACAGGCCAGACTGCAAGGCGAGGATTGCCTTTCTCCAAGGAAGCCCGTCTTCCC
TAGGGTGTGGTACTACTAGGCTACAGCAGCCATCACCTTTATAAAAACCCAAGCCATATTTG
AATTGTCTTTAAATTTCAAGGAAGAGGACAAAGAAATGTGAGCAAGGCTTCATTCCTAGCTCT
CAGACTACCGTCCCAAGAGAACCTGAGGTTCCATGGAGCAGGAATAAAGAGGAGTCCTTACTT
CGTCTGGGGAACACAGATGGGGTATGGGTTCTAAGAAACATTCCCGACCGGT-3'

iCre internal probe Southern blot, length: 798bp
Restriction sites used: XmaI and NcoI

Sequence:
5'-GTGCCCAAGAAGAAGAGGAAAGTCTCCAACCTGCTGACTGTGCACCAAAACCTGCCTGCC
CTCCCTGTGGATGCCACCTCTGATGAAGTCAGGAAGAACCTGATGGACATGTTCAGGGACAGG
CAGGCCTTCTCTGAACACACCTGGAAGATGCTCCTGTCTGTGTGCAGATCCTGGGCTGCCTGG
CAAGCTGAACAACAGGAAATGGTTCCCTGCTGAACCTGAGGATGTGAGGGACTACCTCCTG
TACCTGCAAGCCAGAGGCCTGGCTGTGAAGACCATCCAACAGCACCTGGGCCAGCTCAACATG
CTGCACAGGAGATCTGGCCTGCCTCGCCCTTCTGACTCCAATGCTGTGTCCCTGGTGATGAG
GAGAATCAGAAAGGAGAATGTGGATGCTGGGGAGAGAGCCAAGCAGGCCCTGGCCTTTGAACG
CACTGACTTTGACCAAGTCAGATCCCTGATGGAGAACTCTGACAGATGCCAGGACATCAGG
AACCTGGCCTTCCTGGGCATTGCCTACAACACCCTGCTGCGCATTGCCGAAATTGCCAGAATCA
GAGTGAAGGACATCTCCCGCACCGATGGTGGGAGAATGCTGATCCACATTGGCAGGACCAAGAC
CCTGGTGTCCACAGCTGGTGTGGAGAAGGCCCTGTCCCTGGGGGTTACCAAGCTGGTGGAGA
GATGGATCTCTGTGTCGGTGTGGCTGATGACCCCAACAACTACCTGTTCTGCCGGGTCAGAAAG
AATGGTGTGGCTGCCCCTTCTGCCACCTCCCAACTGTCCACCCGG-3'

6.1.10. Molecular weight markers

DNA ladder:	100bp ladder; 1kb ladder (NEB GmbH, Frankfurt a. M.)
Protein ladder:	SeeBlueR Plus2 Pre-Stained Standard (Gibco, Invitrogen™ Cooperation, Carlsbad, CA)
RNA ladder:	RNA ladder high range (Fermentas GmbH, St. Leon-Rot)

6.1.11. Bacteria and culture media

Bacteria

E. coli K12 EL350	(Lee et al., 2001)
E. coli K12 EL250	(Lee et al., 2001)
E. coli K12 XL-1 Blue	endA1 gyrA96(nalR) thi-1 recA1 relA1 lac glnV44 F'[::Tn10 proAB+ lacIq Δ(lacZ)M15] hsdR17(rK- mK+) (Stratagene, La Jolla)
E. coli K12 DH5α:	F-, acl- recA1, endA1, D(lacZY A-argF), U169, F80dlacZDM15, supE44, thi-1, gyrA96, relA1 (Hanahan et al., 1985)

Culture media

LB medium	(lysogeny broth; Bertani, 1951)
LB agar	(lysogeny broth; Bertani, 1951)
supplemented with	100µg/ml ampicillin
	25 µg/ml kanamycin
	12 5µg/ml chloramphenicol

6.1.12. Cell lines, culture media and solutions

Cell lines

HEK293T	human embryonic kidney cell line, derivative of HEK293 that stably express the T-large antigen of SV40 (Simian Virus 40) (Graham et al., 1977)
NIH3T3	murine embryonic fibroblast cell line (Todaro and Green, 1963)
IDG3.2	murine ES cell line (F1); background: 129Sv/C57Bl/6
TBV2	murine ES cell line; background: 129Sv (Wiles et al., 2000)
MEF	primary murine embryonic fibroblasts, isolated E13.0

Culture media

HEK293T/NIH3T3	DMEM (Gibco, Invitrogen™ Cooperation, Carlsbad, CA), supplemented with 2mM L-glutamine (200mM Gibco, Invitrogen™ Cooperation, Carlsbad, CA), 10% FCS (PAA Laboratories Gesellschaft mbH, Pasching, Österreich)
MEF	DMEM (Gibco, Invitrogen™ Cooperation, Carlsbad, CA), supplemented with 2mM L-glutamine (200mM Gibco, Invitrogen™ Cooperation, Carlsbad, CA), 15% FCS (PAN Biotech GmbH, Aidenbach), 0.1mM β-mercaptoethanol (50mM, Gibco, Invitrogen™ Cooperation, Carlsbad, CA), 1x MEM (non-essentiell amino acids, 100x; Gibco, Invitrogen™ Cooperation, Carlsbad, CA)
TBV2	DMEM (Gibco, Invitrogen™ Cooperation, Carlsbad, CA), supplemented with 2mM L-glutamine (200mM Gibco, Invitrogen™ Cooperation, Carlsbad, CA), 15% FCS (PAN, Biotech GmbH, Aidenbach), 0.1mM β-mercaptoethanol (50mM, Gibco, Invitrogen™ Cooperation, Carlsbad, CA), ESGRO® (LIF) (10⁷ U/ml; Chemicon, Millipore, Schwalbach), 1x MEM (non-essentiell amino acids, 100x; Gibco, Invitrogen™ Cooperation, Carlsbad, CA)
IDG3.2	DMEM (Gibco, Invitrogen™ Cooperation, Carlsbad, CA), supplemented with 2mM L-glutamine (200x, Gibco, Invitrogen™ Cooperation, Carlsbad, CA), 15% FCS (PAN Biotech GmbH, Aidenbach), 0.1mM 3-mercaptoethanol (50mM, Gibco, Invitrogen™ Cooperation, Carlsbad, CA), ESGRO® (LIF) (10⁷U/ml; Chemicon, Millipore, Schwalbach), 1x MEM (non-essential amino acids, 100x; Gibco, Invitrogen™ Cooperation, Carlsbad, CA), 2mM HEPES (200mM, Gibco, Invitrogen™ Cooperation, Carlsbad CA)

Differentiation medium	SFO3 supplemented with 10ng/ml human activin A (0.05mM; R&D Systems GmbH, Wiesbaden-Nordenstadt) 50µl β-mercaptoethanol per 50ml medium (50mM; Gibco, Invitrogen™ Cooperation, Carlsbad, CA)
for selection supplemented with	1-2µg/ml puromycin
	300µg/ml G418 (50mg/ml Geneticin; Gibco, Invitrogen™ Cooperation, Carlsbad, CA)
Solutions for cell culture	
1x PBS without Mg^{2+}/Ca^{2+}	(Gibco, Invitrogen™ Cooperation, Carlsbad, CA)
1x trypsin-EDTA	(0.05 % Trypsin, 0.53 mM EDTA•4Na, Gibco, Invitrogen™ Cooperation, Carlsbad, CA)

6.1.13. Mouse lines

C57Bl/6 N	inbred strain
CD1	outbred strain
$Foxa2^{iCre\Delta neo/+}$	(Uetzmann et al., 2008)
$Sox17^{iCre\Delta neo/+}$	(Liao et al., 2009)
Flp-e	(Dymecki, 1996)
$R26^{R/R}$	(Soriano, 1999)
$\beta\text{-}catenin^{flox/flox}$; $R26^{R/R}$	(Brault et al., 2001; Soriano, 1999)
$\beta\text{-}catenin^{floxdel/+}$	(Brault et al., 2001)

6.2. Methods

6.2.1. Methods in molecular biology

6.2.1.1. Preparations of nucleic acids: DNA preparations

a) Plasmid and BAC preparations

A plasmid preparation, refers to the isolation of plasmid DNA from a bacterial suspension. The bacterial cells have to be separated from the medium first and be solubilized by addition of lysis buffers.
A higher osmolarity of the medium outside and the presence of chelating agents which bind metal ions from the cell wall make the cells porous. Detergents for the lysis of the cell membrane and DNA denaturing solutions are then added. Afterwards the whole medium is neutralized, allowing for the activity of added RNases and the renaturing of the DNA. One is thereby taking advantage of the fact that plasmid DNA renatures quicker than chromosomal DNA due to the plasmids' smaller size. The plasmid DNA stays in solution, while everything else is pelleted by centrifugation. Afterwards plasmid DNA can be precipitated by adding isopropanol, washed with 70% ethanol and resuspended in water.
For longer storage the DNA can be resuspended in TE buffer and stored at -20°C.

I. Plasmid preparations according to the QIAGEN Plasmid Kits
   Plasmid preparations were carried out using the QIAGEN Mini Kit and the DNA pellet was resuspended in up to 50µl dist. H_2O, TE buffer or EB buffer. For larger amounts of DNA the QIAGEN Maxi Kit was used and the DNA was resuspended in 150-300µl TE buffer.

II. BAC mini preparation according to Copeland
   The preparation of BAC DNA is similar to the preparation of plasmid DNA. The protocol was adopted from an existing protocol (Warming et al., 2005). The obtained yield of BAC DNA lies around 1-1.5µg.

   5ml of a bacterial ON culture (LB medium supplemented with 25µg/ml chloramphenicol) were centrifuged at 5000rpm in a 15ml Falcon tube for 5min (centrifuge: 5804 R; Eppendorf). The supernatant was discarded.
   The pellet was resuspended in 250µl P1 buffer and transferred to an Eppendorf reaction tube. 250µl of P2 buffer were added and the reaction tube was carefully inverted. Afterwards an incubation at RT for a maximum of 5min was performed.
   After the incubation, 250µl of P3 buffer were added and the tube was subsequently incubated on ice for 5min.

Material and Methods

The protein precipitate was collected by centrifugation at 13.500rpm for 5min (centrifuge: 5417 R; Eppendorf) and the DNA containing supernatant was transferred to a new Eppendorf reaction tube. This procedure was repeated to eliminate the precipitate completely.
To precipitate the BAC DNA 750µl of isopropanol were added to the clean supernatant, mixed and incubated on ice for 10min. After the incubation the DNA was pelleted at 13.000rpm for 10min (centrifuge: 5417 R; Eppendorf), the supernatant was removed and the DNA pellet was washed using 1m 70% ethanol. Following this step, the DNA was centrifuged again at 13.200rpm for 5min.
The ethanol was removed carefully and the pellet was air-dried up to 10min and then dissolved in 50µl TE by incubation at 37°C and 500rpm (Thermomixer comfort; Eppendorf) for 1h.
Subsequent restriction digests were carried out in a volume of at least 60µl.

III. BAC maxi preparation according to the NucleoBond BAC Purification Maxi Kit

Maxi preparations of BAC DNA were done according to the NucleoBond BAC Purification Maxi Kit. The BAC DNA pellet was resuspended in a suitable volume of TE buffer.

b) Preparation of genomic DNA from cells or tissue

The preparation of genomic DNA from cells or tissue, in principle, consists of three steps: the complete lysis of the cells, the precipitation of the DNA and its subsequent purification.
The lysis is carried out using SDS-containing buffers, which is freshly supplemented with proteinase K to remove proteins, especially DNA associated proteins. The incubation is done at 55°C, the optimal temperature for the proteinase activity.
By adding salt and ethanol to the solution after complete lysis the DNA is precipitated.
Afterwards the DNA is washed with ethanol, dissolved in water or buffer and, if necessary, purified by an additional phenol-chloroform extraction. This last cleaning step is only necessary if following restriction digests with especially sensitive restriction enzymes are needed.

I. Isolation of genomic DNA from cells in 96-well plate format

Comment:
The purification of DNA with phenol-chloroform can be disregarded in 96-well format because of the effort. Therefore the following restriction digests should be tested with DNA prepared according to the same protocol. To obtain enough material the extraction of DNA from cells in 96-well format should be started when the cells are confluent and the medium (100µl/well) turns yellow or orange within one day.

Directly before starting the preparation the lysis buffer was supplemented with proteinase K in a concentration of 100µg/ml, because the proteinase will digest itself over time.
The cells were washed twice with PBS–Mg^{2+}/Ca^{2+} to completely get rid of the medium. The PBS was removed and 50µl of lysis buffer were added to each well. The plate was sealed with parafilm and incubated at 55°C in a humid chamber ON.
The next day 150µl 5M sodium chloride were mixed with 10ml 100% ice-cold ethanol and 100µl of this mixture was added to each well for precipitation. The plate was incubated at RT for 30min without moving.
To decant the liquid, the plate was inverted carefully and slowly (in about one minute). Rests of the liquid were removed by putting it on a paper towel upside-down.
Subsequently, the DNA was washed three times using 150µl 70% ice-cold ethanol per well. For each wash, the plate was inverted as previously described. At this point the DNA can be stored in 70% ethanol at -20°C.
After the last washing step the DNA was dried at RT for 10-15min. Then 25µl TE buffer or water were added to the pellet and the DNA was dissolved at 4°C ON or at 37°C, shaking, for 1h in a humid chamber.

II. Isolation of genomic DNA from cells in cell culture dish format

To identify transgenic ES cell clones, these clones were expanded to one 10cm dish and DNA was isolated yet again, independently from the first results obtained with DNA from 96-wells.

First the medium was removed from the cells and they were washed once with PBS (-Mg^{2+}/Ca^{2+}).
The lysis buffer was supplemented freshly with proteinase K to a final concentration of 100µg/ml. 5ml of this solution were used for lysis per 10cm dish.
For this purpose the lysis buffer was pipetted onto the cells, the cells picked from the dish using a cell scraper and transferred to a 50ml Falcon tube. Afterwards the reaction took place at 55°C ON.
Phenol-chloroform-isoamylalcohol (25:24:1, PCI) was added 1:1 to the cell solution that was incubated ON and mixed well by vortexing.
After that the phases were separated by centrifugation for 10min at 4500xg and RT.
The upper phase was removed without the protein-containing interphase and mixed 1:1 with PCI again – the

phases separated the same way. The upper phase was now transferred to a new tube and the DNA precipitated by adding 10ml ethanol:sodium-acetate (25:1). Subsequently, the DNA was pelleted by centrifugation at 4500xg and 4°C for 10min. The supernatant was discarded and the pellet was washed with at least 7,5ml 70% ice-cold ethanol. Centrifugation was carried out at 4500xg and 4°C for 5min. The supernatant was discarded again and the DNA pellet was dried at RT until it appeared glassy. After that, the pellet was dissolved in a suitable volume TE buffer at 4°C ON or at 37°C, shaking, for 1h and stored at 4°C.

III. Isolation of genomic DNA from mouse tail biopsy

The preparation of genomic DNA from mouse tail tips provides the basis for genotyping the mice. For efficiency reasons the genotyping relies on the principle of the PCR and therefore it is important that preparations are done as clean and carefully as possible to prevent contamination.

Each individual mouse tail tip with a length of approximately 4mm was put into a single Eppendorf reaction tube and was stored at -20°C if it could not be processed the same day.

The lysis buffer for mouse tail tip DNA was freshly supplemented with proteinase K (final concentration: 100µg/ml) and each mouse tail tip was lysed in 500µl of this buffer at 55°C ON.

The next day the complete lysis was visually checked and the suspension mixed properly by vortexing. By centrifugation at 14.000rpm (centrifuge: 5417 C; Eppendorf) for 10min the mouse hairs and the other remaining insoluble constituents were pelleted. The supernatant containing the DNA was transferred into a new Eppendorf reaction tube and containing 500µl isopropanol. To precipitate the DNA completely everything was mixed well by stringently shaking. Afterwards the DNA was pelleted at 14.000rpm (centrifuge: 5417 C; Eppendorf) for 20min, the supernatant was decanted and discarded, the DNA was washed once with 70% ethanol (centrifugation as before for 5min) and the pellet was air-dried at RT for 10-15min. Finally the pellet was dissolved in up to 500µl H_2O or TE buffer and incubated at 4°C ON or at 37°C, shaking, for 1h.

6.2.1.2. Preparations of nucleic acids: RNA preparations

RNA preparations are carried out analogously to DNA preparations and can therefore be divided into three basic steps: complete lysis of the cells, precipitation and purification of the RNA.

During the RNA extraction it is important to work as fast as possible and especially RNase-free. The use of RNase inhibitors is highly recommended. For instance, cells can be lysed using buffers containing β-mercaptoethanol; a chemical solubilizes the cells and inhibits the activity of RNases

RNA should be dissolved in diethylpyrocarbonate (DEPC) treated buffers (MOPS or PBS) or in DEPC H_2O. DEPC inactivates RNases irreversible by destroying their histidine residues. DEPC, however, is non-compatible with tris or HEPES buffers and should not be used in those cases. Extracted RNA is most stable when stored at -80°C.

a) Preparation of total RNA according to the QIAGEN RNA Mini Kit

Total RNA preparations were carried out using the QIAGEN RNA Mini Kit. RNA was stored at -80°C.

6.2.2. Determination of the concentration of DNA and RNA solutions

The concentrations of nucleic acid solutions were determined measuring the extinction at 260nm with a photometer (NanoDrop). This specialized photometer allows the measurement of concentrations with extremely small amounts of solution (1µl).

The concentration of genomic DNA in solution, however, cannot be determined this way. Concentrations of this kind of DNA have to be estimated by agarose gel electrophoresis.

The correlation between extinction absorption and concentration is displayed by the following equation (law of Lambert-Beer):

$$c (RNA/DNA) = E_{260nm} / (d \times \varepsilon) \times V$$

c: concentration [µg/ml]
E_{260nm}: measured extinction (absorption) at 260nm
d: layer thickness of the cuvette [cm]
V: dilution factor
ε: extinction coefficient (depending on the kind of nucleic acid, s. b.) [ml/(cm x µg)]

ε_{dsDNA} = 1/50 ml/(cm x µg)
ε_{ssDNA} = 1/40 ml/(cm x µg)
ε_{RNA} = 1/40 ml/(cm x µg)

Material and Methods

To control the purity of the nucleic acid solution the extinction at 280nm (band of absorption by proteins) was determined.
The quotient E_{260nm}/E_{280nm} should be between 1.9-2.0 for clean RNA solutions and between 1.8-1.9 for clean DNA solutions.

6.2.3. Reverse transcription

a) DNase digest of RNA samples

To remove remaining DNA in the RNA samples it is necessary to digest the RNA with DNase.
A DNase digest was carried out as follows:

11µg	RNA
1.0µl	RNAse inhibitor (40U/µl > 40U)
6.0µl	RQ1 DNase (6U/µl > 6U)
3.0µl	RQ1 10x DNase buffer (1x)
30.0µl	final volume with DEPC treated H_2O

The mixture was incubated at 37°C for 1h. To stop the reaction approximately 10% (3µl) RQ1 DNase Stop Solution were added and incubated at 65°C for 10min.

b) RNA precipitation with LiCl

After DNAse digest it is necessary to remove all protein and chemicals before the reverse transcription is carried out to allow for a proper transcription.
To precipitate the RNA the mixture used for the DNase-Verdau was filled up to 100µl final volume with DEPC treated H_2O. Then 10µl 4M LiCl and 300µl 96% ethanol were added (threefold volume). The mixture was incubated at -80°C for 30min to precipitate the RNA. Subsequently it was centrifuged at 4°C and 14.000rpm (5417 R, Eppendorf) for 10min. The supernatant was discarded and the RNA pellet was washed with 200µl 70% ethanol (in DEPC treated H_2O). Again RNA was centrifuged under the same conditions for 5min. Afterwards the pellet was dried till it appeared glassy and finally it was resuspended in 10µl DEPC treated H_2O.

c) Reverse transcription with oligo dT-primers

In a reverse transcription RNA is transcribed into DNA (cDNA). Using oligo-dT-primer polyadenylated mRNA is transcribed. For each RNA sample one preparation with and one without reverse transcriptase should be made, especially when the primers used in the following PCR do not distinguish genomic DNA and cDNA (without intron), means that the primers are not intron-spanning.
The following mixture was used for one cDNA preparation:

2.0µl	RNA (from DNAse digest > 2µg)
1.0µl	Oligo-dT primer (500ng/µl)
11.0µl	filled up with DEPC-treated H_2O

The preparations were incubated at 70°C for 10min to denature the RNA and allow for the annealing of the primers. Afterwards the mixture was put on ice and the following components were added:

4.0µl	5x transcriptase buffer
2.0µl	DTT (0,1M > 10mM)
1.0µl	dNTPs (je 10mM > 0,8mM)
1.0µl	RNA inhibitor (40U/µl > 40U)

The mixture was incubated at 42°C for 2min. Then 200U SuperScript (200U/µl > 1µl) was added and subsequently it was incubated at 42°C for 50min. To stop the reaction the enzyme was inactivated at 70°C for 15min. The cDNA was stored at -20°C.

d) PCR on cDNA

For the detection of gene transcription PCR on cDNA was carried out as follows.

Material and Methods

mixture		programm:		
0.5µl	cDNA	94°C	5min	
2.0µl	10x buffer			
1.0µl	MgCl$_2$	94°C	1min	
2.0µl	primer sense (10pM)	XX°C	30s	} 30-40x
2.0µl	primer antisense (10pM)	72°C	45s	
2.0µl	dNTPs (10mM)			
0.5µl	Taq (Fermentas GmbH)	72°C	10min	
10.0µl	dist. H$_2$O	4°C		
20.0µl	Σ			

6.2.4. Restriction analysis of DNA

Restriction analyses for characterization of DNA are based on enzymes, which are able to cut DNA. Originally they were a defensive mechanism of bacteria against foreign DNA (e.g. DNA of phages).
They recognize palindromic sequences on the DNA strand and they cut the DNA either approximately 100 nucleotides after the recognition site (restriction endonucleases class 1), directly at the recognition sequence (restriction endonucleases class 2) or at a defined distance from the recognition site (restriction endonucleases class 3).
After cutting, restriction enzymes can leave DNA fragments with 3´ or 5´ overhang (sticky ends) or plain ends without any overhang (blunt ends).
If the concentration of a restriction enzyme in one reaction exceeds a certain level (>5-10% of the reaction volume) the risk of STAR activity, unspecific cutting by the enzyme, may occur. With the decrease of the salt concentration of the solution the specificity of the enzyme decreases also; concentrations which are too high can eventually lead to a failure of the reaction.

a) Analytical or preparative restriction digest of plasmid or BAC DNA

A typical reaction for a restriction digest for plasmid or BAC DNA is as follows:

DNA from mini preparation (ca. 500-100ng)	5.0µl
10x buffer	2.0µl
10x BSA (optional)	2.0µl
enzyme (5.000U)	0.5µl
dist. H$_2$O	10.5µl
Σ	20.0µl

The digest was incubated in 1.5ml Eppendorf tubes at 37°C or 25°C for 1.5h.

b) Restriction digests of genomic DNA

In principle, the same rules are valid for restriction digests of genomic DNA used for Southern blot analysis as for restriction analysis of plasmid DNA. The suitable enzyme buffer and BSA, if necessary, are added to the reaction. RNaseH can also be used to get rid of RNA background and thereby raise the efficiency of the enzyme activity. Additionally spermidine (in a final concentration up to 2mM) makes sure that the digest will be complete.
The completeness of the digest is controlled visually on the gel before blotting. The DNA should appear on a lower molecular level and repetitive sequences should emerge in a distinct band.

I. Restriction digests of genomic DNA from (ES-) cells and tissue for Southern blot analysis

A typical mix for a restriction digest of genomic DNA was carried out as follows:

25µl genomic DNA
1x restriction buffer
(1x BSA)
1x spermidine (1mM)
50-100µg/ml RNaseH
H$_2$O to fill up
approx. 50U restriction enzyme

The reaction was incubated at the enzyme-dependent optimal temperature ON. Prior to loading, samples were partially evaporated to obtain a smaller volume.

II. Restriction digests of genomic DNA from (ES-) cells in 96-well plates for Southern blot analysis

A typical mix for a restriction digest of genomic DNA in 96-well plates was carried out as follows:

25.00µl DNA (whole DNA preparation of one well)	
0.40µl 100x BSA	c_{end} = 1x BSA
0.40µl 100mM spermidine	c_{end} = 1mM spermidine
0.25µl RNase A 1mg/ml	c_{end} = 6.25µg RNase
4.00µl 10x enzyme buffer	c_{end} = 1x enzyme buffer
2.50µl enzyme (10U/µl)	c_{end} = 25U enzyme/reaction
7.45µl H$_2$O	
40.00µl	

A master mix of all components except DNA was added to the 96-well DNA using a Multipette (Eppendorf) or an 8-channel pipette (Eppendorf) and incubated at 37°C or 25°C (depending on the optimal temperature of the enzyme) ON. Due to the use of small volumes and poorly closing plates, a humid chamber was used to prevent evaporation and drying of the samples. In principle, this digest could be carried out in a larger volume and after incubation the volume could be reduced via evaporation at 37°C with slightly opened lid to make gel loading of the whole sample possible.

6.2.5. Gelelectrophoresis

Gelelectrophoresis is a tool for the analysis of the size of molecules, e. g. DNA, RNA and proteins.
The gel forms a fine-pored net depending on the percentage of the agarose (for DNA and RNA) or polyacrylamide (PAA; for proteins). The sample is transported through this net using electricity. The velocity of the sample is correlated with the size of the molecule and its three-dimensional shape (ideal case: linear), its net charge, the power of the electric field, the pore size of the gel and the temperature.

a) Analytical agarose gelelectrophoresis

I. DNA

A typical analytical agarose gel was prepared according to expected fragment size with 0.8-2.0% agarose in TAE buffer. The agarose was dissolved by heating in the microwave. After the solution was cooled down, 5µl ethidiumbromide (EtBr)-solution (c_{EtBr}=1mg/µl, end concentration: 0.005%) were added, mixed by swaying and poured into a prepared gel chamber without bubbles.
Separation of samples from PCR or restriction analysis was carried out at U=100V for ca. 30-45min. The band patterns of the samples were documented under UV light.

II. RNA

The integrity of RNA was proven and documented on 1.0% agarose gels in TAE including EtBr analogously to DNA agarose gels using a voltage of U=120V for 10min.

b) Preparative agarose gelelectrophoresis

Preparative gel electrophoresis was carried out analogously to the analytical gel electrophoresis, but the bands with the right size were cut out and documented under UV with less intensity to avoid UV-light induced mutations.

6.2.6. Generation of blunt ends

For cloning it might sometimes be the only way to use two enzymes which overhangs are not compatible. Because not-compatible sticky DNA ends cannot be ligated, these ends have to be converted to blunt ends. There are in principle two ways to achieve blunt ends from sticky ends: either the strand is filled up (for 5'-overhangs) or it is blunted by cutting of the overhanging bases (for 3'-overhangs) using Klenow fragment.
Typical mixes for blunting using Klenow fragment are shown.

```
 5.0µg  DNA
 1.0µl  Klenow fragment (5U/µl)
 1.0µl  10x buffer suppl. with 33µM dNTPs
 1.0µl  ATP (10mM)
 2.0µl  H₂O
10.0µl
```

6.2.7. Dephosphorylation of linearised DNA

For cloning it might be necessary to dephosphorylate especially vector DNA if the overhangs left by the restriction enzymes after digest are compatible. Otherwise the efficiency for the ligation of the insert into the vector goes down because religation of the vector will be energeticly favoured.

A typical mix for the dephosphorylation of linearized plasmid DNA carried out as follows:

```
 1.0µg  DNA
 1.0µl  10x enzyme buffer                      c_end = 1x enzyme buffer
 1.0µl  alkaline phosphatase (1U/µl)           c_end = 1U enzyme/reaction (1µg DNA)
 7.0µl  H₂O
10.0µl  add at
```

6.2.8. Ligation

Ligations of DNA fragments can be done with compatible ends left over by restriction enzymes. It does not matter if these overhangs are 3' or 5' or if there is a blunt end.
T4 DNA ligase is an enzyme extracted from the phage T4 which can ATP-dependently ligate DNA fragments with at least one phoshorylated 5' end. The buffer supplied with enzyme contains 10mM ATP and should therefore be kept frozen.
Vector and Insert should be used in a ratio 1:3 for best results in a ligation with sticky DNA ends. In a blunt end ligation a ratio 1:1 vector:insert is recommended.

A typical mix for the ligation of vector with insert is the following:

```
 1.0µl  vector DNA (e.g.; see below)
 0.5µl  insert DNA (e.g.; see below)
 1.0µl  10x T4 ligation buffer                 c_end = 1x enzyme buffer
 0.5µl  T4 ligase (10U/µl ; NEB)               c_end = 5U enzyme/reaction
 7.0µl  dist. H₂O (add at 10µl with dist. H₂O)
10.0µl
```

The used volume of insert can be calculated with the following equation:

Ratio: vector/insert = 1/3

vector concentration:	$c(v)$ [ng/µl] = x ng/µl	size (v) [bp] = y bp
insert concentration:	$c(i)$ [ng/µl] = a ng/µl	size (i) [bp] = b bp
used amount of vector-DNA:	100-400ng	> used for ligation: v ng / w µl
used amount of insert-DNA:	intron [ng] = 3 * (b bp / y bp) * w µl	> intron DNA [µl]

6.2.9. Cloning of short DNA sequences using complementary DNA oligonucleotides

When cloning short DNA fragments it is sometimes easier to avoid complicated DNA amplification using PCR and rather rely on the use of oligonucleotides, which are available for purchase.
In principle oligonucleotides are used that show the desired sequence and that are complementary to each other. That means that the DNA double strand is generated by simple alignment of the two DNA fragments.
By picking appropriate nucleotide sequences for the 5' and 3' ends of the nucleotide chains, overhangs can be created that mimic restriction sites, which will simplify the cloning considerably. Moreover, an additional phosphorylation of the resulting double stranded fragments can enhance the ligation efficiency.

a) Hybridisation

The following components were mixed for complementary hybridisation of oligonucleotides:

94.0µl	dist. H_2O
2.0µl	1M Tris, pH 7.5
0.5µl	0.5M EDTA, pH 8.0
50.0µl	5´>3´ oligonucleotide (100pmol/µl)
50.0µl	3´>5´ oligonucleotide (100pmol/µl)

The following PCR program was used to first denature (abolishment of secondary structures) and secondly slowly renature the DNA strands to guarantee a precise and accurate complementary attachment:

100°C	10min
65°C	20min
50°C	20min
37°C	20min
RT	∞

b) Phosphorylation

Thereafter the double strand fragments were phosphorylated for the proximate ligation using the following approach:

2.0µl	mixture (see above), hybridised oligonucleotides
1.0µl	10x T4 polynucleotidkinase (PNK) forward buffer
1.0µl	10mM ATP (final concentration: 1mM)
5.0µl	dist. H_2O
1.0µl	PNK (Fermentas GmbH)
10.0µl	Σ

The program for the phosphorylation reaction was as follows:

37°C	1h	(phosphorylation)
70°C	10min	(inactivation)

c) Ligation of hybridised oligonucleotides

Subsequent to the phosphorylation reaction the ligation was started using the following approach:

1.0µl	10x ligation buffer
2.0µl	vector DNA (5µg digested ON, dephosphorylated where necessary, purified with the Qiagen PCR-Purification Kit)
0.5µl	hybridised oligonucleotides (from phosphorylation reaction, see above)
0.5µl	T4 DNA ligase (NEB)
6.0µl	dist. H_2O
10.0µl	Σ

The ligation was performed as follows:

16°C ON.

6.2.10. Generation of competent bacteria

a) Generation of electro-competent bacteria (E. coli K-12 XL1-Blue)

recA1 endA1 gyrA96 thi-1 hsdR17 supE44 relA1 lac [F´ proAB lacIqZΔM15 Tn10 (Tetr)]

Out of a 2ml pre-culture bacterias were plated on LB plates with kanamycin (30µg/ml), tetracyclin, ampicillin (40µg/ml) and with antibiotics as controls and incubated at 37°C over night.
The following day, if the control plates have proven that there cannot be found any adapted resistance, a 50ml preculture is set up for the next day using a single colony and incubated over night at 37°C while shaking.
All material that will be used should be washed carefully with clean water. Any detergent that is laft will reduce the efficiency of competence. The following day 500ml LB medium are inoculated with 5ml bacterial preculture at 37°C until OD_{600} = 0.5 (~3h).

The following steps are all carried out on ice or at 4°C to prevent the bacterias from warming. Every material used should be precooled. First the bottles with the bacterial suspension were put in ice water for at least 15min to cool them down while swirling every few minutes. The bacteria were then pelleted in clean centrifugation bottles (2x 250ml) by spinning them for 15min at 4000xg in the precooled centrifuge. The pellets were resuspended in 250ml ice-cold water each and spinned again at 6000xg for 15min. Increasing g is necessary to avoid losing a lot of bacteria. After centrifugation the each pellet was again resuspended in 125ml ice-cold water and spinned under the same conditions (6000xg; 15min; pre-cooled). After this centrifugation step the bacteria were resuspended in 5ml pre-cooled 10% glycerol solution per pellet and transferred to pre-cooled 30ml Correx-tubes. The cells were then centrifuged at 8000xg for 15min. Finally, the pellets were resuspended in 0.5ml fresh pre-cooled 10% glycerol solution. Aliquots à 40µl were made and fastly frozen down on dry ice. The bacteria were kept on –80°C.

6.2.11. Transformation of bacteria

The term transformation marks the transfer of foreign DNA e. g. a plasmid or BAC (Bacterial Artificial Chromosome) into organisms, e. g. bacteria. For this kind of manipulation of bacteria there are in principle two common methods: chemical transformation and electroporation.

The chemical transformation is based on the assumption that the membrane gets porous and instable through Ca^{2+} ions and heat, so that the pick up of DNA sticking to the membrane is possible.

Using electroporation, another method of transformation of bacteria and eukaryotic cells, the cells are made permeable for plasmids or other vectors in a salt-free solution by shortly applying a high electric tension. This method allows an efficient transfer of big DNA fragments into diverse organisms but by which the vitality certainly is relatively low.

Independent of the kind of transformation the bacteria or cells should have a regeneration time of a few cell cycles (generally 1-3 cycles) in normal medium and under normal incubation conditions. Only then they should be transferred to selection media.

a) Transformation of bacteria using electroporation

One aliquot of electrocompetent bacteria per construct to be transformed was thawed on ice. 1-2µl of DNA for transformation was incubated with the bacteria on ice for 5min. The mixture was transferred into a pre-cooled electroporation cuvette (0.1cm) without any bubbles. Electroporation was carried out with a tension of U=2.5kV. The space of time of the pulse should lie between 4.5 and 5.0ms.

After the pulse the bacteria were immediately transferred into a 1.5ml tube (Eppendorf) using 1ml LB medium. For regeneration the transformed bacteria were kept shaking (850rpm; Eppendorf shaker) at 37°C for 30-60min. Then they were put on LB plates supplemented with appropriate antibiotics in a suitable dilution.

b) Transformation of bacteria using heat shock

For the transformation of bacteria with heat shock, one aliquot of bacteria per construct to be transferred was thawed on ice. A suitable amount (e.g. 4µl of a ligation) of Vector was mixed with the bacteria and incubated on ice for 30min. For heat shock the mixture was put on 42°C in a water bath for 90s. Immediately after the heat shock, the bacteria were put back on ice, diluted with 1ml LB medium and incubated for regeneration (850rp, Eppendorf shaker; 37°C; 30-60min).

6.2.12. Bacterial homologous recombination

2-4ml LB medium were incubated with Bacteria from a glycerol stock of EL-250 / 350 at 32°C over night (ON). The next day the ON culture was diluted 1:50–1:100 and grown in 50ml LB medium to OD_{600}=0.6

Then the culture was split: 25ml were left in a Falcon tube at room temperature (recombineering control), the other 25ml were put in a water bath at 42°C while shaking for 15 min. Afterwards both cultures were cooled down on ice for 5min and kept cool from now on.

Bacteria were centrifuged in 50ml Falcon tubes at 5000rpm and 4°C for 5-10min (Eppendorf 5804 R).

Then the bacterial pellets were resuspended in 1.8ml of ice-cold 10% glycerol each and transferred into 2ml tubes. The suspensions were again centrifuged at full speed (14000rpm, Eppendorf 5417 R) and 4°C for 20s and again resuspended in 1.8 ml 10% ice-cold glycerol (vortexing), This washing step was repeated three more times.

After the last washing step the pellet was resuspended in 10% glycerol to a total volume of 100µl – 50µl of the bacterial suspension were used for each electroporation.

Two electroporations were prepared (heat-shocked and non-heat-shocked bacteria). 15ng of the linearised PL-254 vector with retrieval PCRs (e.g.) was mixed with 50µl of competent cells, left on ice for 5min and was then electroporated.

Material and Methods

The cells were grown at 32°C in 1.5ml tubes for 30min, then plated in two dilutions on LB+selection plates: 200µl (1:5) and 20µl (1:50). The plates were incubated at 32°C ON.

6.2.13. DNA sequencing

The current method of DNA sequencing is based on the principle of chain termination, originally published by the Nobel laureate Frederick Sanger in 1977.
Natural DNA synthesis lacks the ability to start de novo, but rather has to start with a short DNA fragment, a so-called primer or oligonucleotide. Nucleotides can then be added to the 3'-hydroxy group of the last free nucleotide, complementary to the anti-parallel strand.
In the Sanger method, labelled dideoxynucleotides, which are missing the reactive OH-group at the 3'-position, are added to the reaction in catalytic amounts. Every time a dideoxynucleotide is attached to the growing DNA strand, the reaction terminates at that exact position. This way a mixture of DNA fragments with different molecular weights is obtained, from which one can identify the last nucleotide due to the labelling.
Subsequently, the fragments are separated according to size using gel electrophoresis or chromatography. The result is a series of labelled nucleotides from which the sequence of the DNA fragment can be read directly. Each individual sequencing reaction allows one to read several hundred base pairs.

a) Sequencing reaction

The following mixture is needed for a sequencing reaction:

long sequence	short sequence	
1.0µl	0.5µl	Big Dye (contains polymerase)
1.0µl	2.0µl	Big Dye – buffer
10pM	10pM	primer (sense/ antisense)
(n) bp/100 = x ng		template-DNA
∑ 5.0µl		

The PCR program for the sequencing reaction was programmed as follows:

96°C	1min	
96°C	10s	⎫
50°C	5s	⎬ 35x
60°C	4min	⎭
4°C	∞	

b) DNA preparation for sequencing: ethanol – sodium acetate precipitation

In a 70% ethanol and sodium acetate solution, DNA precipitates and can be collected by centrifugation, thereby eliminating undesired elements of previous reactions, e.g. proteins. For the sequencing reaction, using the ABI 3730 DNA Analyser (PE Applied Biosystems) sequencing machine, only clean DNA in HPLC purified water should be used to avoid damaging the machine as well as to guarantee a proper analysis

For precipitation of DNA in a 5.0µl sequencing PCR reaction (s. 6.2.12)

    0.5µl 125mM EDTA
    2.0µl 3M sodium acetate and
    50µl 100% ethanol

were added to the PCR reaction solution, mixed well and incubated at RT for 15min. Afterwards the DNA was pelleted by centrifugation at 11.000rpm at 4°C (Eppendorf Zentrifuge 5417) for 30min. The DNA was then washed with 70µl 70% ethanol and centrifuged under the same conditions for 10min. Subsequently, the ethanol was removed, the DNA was air-dried and the pellet dissolved in 15-25µl HPLC-H_2O.
The dissolved DNA was transferred to a 96-well plate. Empty wells were filled with the same volume of HPLC-H_2O and only every odd column of wells was filled with samples (+) and every even column with HPLC-H_2O for flushing the sequencer (see filling scheme).

Filling scheme:

Material and Methods

	1	2	3	4	5	6	7	8...	
A	+	H₂O	+	H₂O	+	H₂O	+	H₂O	
B	+	H₂O	+	H₂O	+	H₂O	+	H₂O	
C	+	H₂O	+	H₂O	+	H₂O	+	H₂O	etc.
...									

6.2.14. Southern Blot

The Southern blot, a method in molecular biology generated by Edwin Southern in 1975, is a common technique to identify special genomic sequences within a complex mixture of DNA (e.g. the whole genome of an individual). The DNA is digested into smaller pieces using restriction enzymes and separated by gel electrophoresis. The pieces of DNA are blotted on a membrane and the fragment of interest is hybridised to a radioactive- or fluorescent-labelled complementary DNA probe. Therefore it can be visualized later by exposing the hybridised membrane to a film.

Additionally the Southern blot can be used to show correct integration during homologous recombination. In this case it is important that the knock-in allele shows another restriction pattern in comparison to the wild type allele. That means that the band, which can be detected with the probe, has to show the change in the fragment size.

a) Gel electrophoresis

First, a photo of the gel was taken under UV light. A ruler was used for reference length, put next to the DNA ladder. If the digest is complete, repetitive sequences can be easily detected by a distinct band under UV light.

b) Blot

For the detection of larger fragments (≥10.000kb), the gel was depurinized by incubating it in 989ml H₂O + 11ml HCl (conc.) for 15-20min while shaking. This was followed by a denaturation step in 0.4M sodium hydroxide and 0.6M sodium chloride again while shaking. For neutralization the gel was incubated in a solution of 0.5M tris and 1M sodium chloride (pH 7.2) for 15-20min once again while shaking. The blot was built up in 20x SSC as follows (see figure 47) and the transfer was carried out ON. Note that the gel lies upside down.

The next day the blot was taken apart and the slots of the gel were marked on the membrane with a ball pen. Afterwards the membrane was dried and cross-linked in between two layers of Whatman 3 MM paper for 30min at 80°C or treated in the cross-linker (UV Stratalinker 1800; Stratagene) for 1min, respectively.

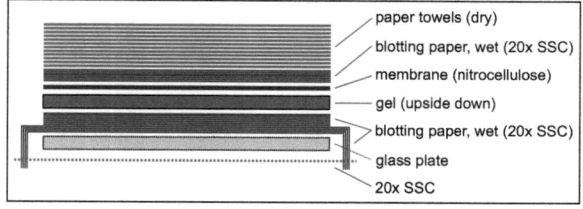

Figure 47: Southern blot setup
The figure schematically shows how the Southern blot was build up in a tray or 20x SSC solution. Indicated are 20x SSC in a tray, glass plate, blotting papers, gel, membrane and paper towels.

c) Hybridisation

I. Prehybridisation
   30ml hybridisation buffer per hybridisation tube (Hybridizer HB 100; ThermoHybaid, Thermo Electron Cooperation) were preheated to 65°C in a water bath. The membrane was rolled and put into the hybridisation tube and the preheated buffer was carefully poured into without generating bubbles between membrane and tube. The membrane was prehybridised turning in an oven at 65°C for at least 1.5 – 2h.

II. Preparation of samples – radioactive labelling of the probe
   While the membrane was prehybridizing, the radioactive labelling of the probe was started. 100ng of the linearised probe DNA was diluted to a final volume of 24µl with dist. H₂O in a 1,5ml Eppendorf reaction tube. 10µl of random oligonucleotides were added and the mixture was boiled for denaturation in a water bath for 5min first.
   Subsequently the tube was put on ice to let the solution cool down, centrifuged to collect the liquid on the bottom of the tube and then put on ice again. 10µl 5x buffer (d*CTP-buffer for 32P labelled CTP) that already incorporates the correct mixture of nucleotides and salts were added. In the radioactive lab labelled dCTP (50Cu) and 5U Klenow-enzyme (Exo(-) Klenow, 1µl, 5U/µl) were added, the mixture carefully vortexed or

Material and Methods

well-mixed, shortly centrifuged and labelled at 37-40°C for 2-10min. For completion of the labelling 2µl STOP-mix were added. This buffer contains the chelating agent EDTA for binding of Mg^{2+} and Ca^{2+} ions and thereby inhibits the activity of the polymerase. For the purification (removal of proteins and residual nucleotides) the reaction mixture was loaded on a prepared column (centrifuged without sample for 2min at 400xg) and centrifuged at 400xg for 2min in a table top centrifuge. Afterwards the sample was put on ice, shortly mixed and centrifuged before measuring and 1µl was used to measure the activity.
For Denaturation of the probe 500µl salmon sperm DNA (10mg/ml) were denatured at 100°C in a water bath for 10min, put in a 50ml Falcon tube and stored on ice. The hot probe was added in a final concentration of $1x10^6$ counts per 1ml hybridisation buffer that means $2x10^7$ counts per 20ml buffer.
By adding 50µl 10N sodium hydroxyl the sample was denatured while the tube was swung carefully to mix. Further on swinging, 300µl 2M tris, pH 8.0, and afterwards 475µl 1M HCl were pipetted drop wise into the tube for neutralisation.

III. Hybridisation

At the end of the prehybridization 10ml hybridization buffer were decanted from the tube and discarded. The radioactive-labelled and chemically-denatured DNA probe was pipetted into the glass hybridization tube and the tube was put back into the oven on 65°C. The incubation was carried out while rotating at 65°C ON (12-24h).

The following day the probe was decanted from the glass tube and the membranes were washed with 2x SSC/ 0.5% SDS (approx. 250ml, RT) in a plastic bowl. For the first washing step the membranes were incubated 5min at RT while shaking.
After 5min the membranes were taken from the SSC solution and the buffer was discarded. 500ml fresh 2x SSC / 0.5% SDS preheated to 65°C were poured into the bowl and the membranes were washed up to 30min while shaking in a 65°C warm water bath. The washing buffer was replaced as necessary. If a new probe was used control measurements were carried out every 5min. Obtaining a signal that was still high but relatively weak (approx. $100-200x10^2$ counts) the washing step was simply repeated. The stringency of the washing steps could be raised by lowering the concentration of SSC (1x SSC / 0.5% SDS or 0.1x SSC / 0.5% SDS, respectively) if higher signals were measured (> $200 x10^2$ counts). Finally, at approx. $35x10^2$ counts (depending on, among other things, the size of the membrane and the probe and its CTP content), the membrane was wrapped tightly in saran wrap and fixed in a film cassette A film (Kocak, BioMax) was applied, stored at −80°C ON and developed the following day (AGFA Curix 60).

d) Preparation of samples - radioactive labelling of the probe according to Roche

Working with reagents of the Roche labeling kit the following changes were applied:
100ng linearised DNA were filled up to 9µl final volume with dist. H_2O and denatured in a boiling water bath for 5min. Afterwards it was shortly put on ice to cool down, centrifuged and put on ice again to prevent renaturation. 1µl dATP, dGTP and dTTP each and 2µ hexanucleotide-mix were added. The addition of radioactive labelled dCTP (50uCi) and Klenow-enzyme was carried out as described above; however, the incubation was done at 37°C for 1h. Terminating the reaction was achieved adding 2µl 0.2M EDTA, pH 8.0, and the probe was diluted to a final volume 50µl using 28µl TE buffer for loading on the column. The remaining steps are as described above.

6.2.15. Detailed description of the generation of the targeting construct for *Sox17*

The knock in construct was generated as follows.
5' and 3' homology regions for the *Sox17* gene were amplified by PCR (primers used for amplification: 5' homology region: fwd 5' HR *NotI*-AgeI, rev 5 HR *HindIII-NdeI*; 3' homology reg on: fwd 3' HR *HindIII*, rev 3' HR *SpeI*; see Material and Methods for sequence of all primers mentioned) using a 129Sv BAC (RPCI22-46A17; Osoegawa *et al.*, 2000) as a template and subcloned into 5'-*NotI* and 3'-*SpeI* sites of the pL254 vector (Lee *et al.*, 2001; Liu *et al.*, 2003; modified as follows: pL253 was cut with *NotI* and filled up with two annealed oligonucleotides: sense EP108, reverse EP109). Using gap repair a 9.3kb genomic fragment was retrieved from the same BAC via homologous recombination in EL350 bacteria as described previously (Lee *et al.*, 2001; Liu *et al.*, 2003; see Material and Methods), resulting in pL254-*Sox17*. *HindIII* and *NdeI* sites were used as single cutters for linearization prior to gap repair.
For cloning of the knock-in cassette in pBluescript KS- (pBKS-) 5' and 3' homology regions for the knock-in into exon 1 of *Sox17* were generated by PCR (5' homology region (HR): EP006 *SacII Sox17* Exon 1A sense, EP004 *NotI Sox17* Exon 1A reverse; 3' homology region: EP009 *EcoRI Sox17* Exon 1B sense, EP010 *HindIII Sox17* Exon 1B reverse) using the above mentioned BAC as a template and subcloned into pBKS- using the introduced restriction sites (underlined), resulting in pBKS- Ex1-HR.
The *iCre* sequence was amplified and subcloned into pBKS- using primers carrying a *NotI* site and a perfect

Material and Methods

Kozak sequence (GCCACC) 5' and a *SpeI* site plus a translational stop codon (TGA) 3' (EP015 *NotI Kozak-iCre* sense, EP016 *SpeI Kozak iCre* reverse) using pBlue.*iCre* (Shimshek *et al.*, 2002) as a template, resulting in a plasmid named pBKS- – *Kozak-iCre*. The artificial intron was then amplified by PCR using pBRY-AM (Fehling *et al.*, 2003) as a template and subcloned into pBKS- – *Kozak-iCre* using *SpeI* and *BamHI* introduced through the 5'- and 3'-primers, respectively (EP011 *SpeI* Intron-pA fwd, EP012 *BamHI* Intron-pA reverse), producing pBKS- – *Kozak-iCre*-Intron-pA.

In the next step the PGK promoter driven neomycin resistance gene flanked by *FRT* sites was cloned from pL451Δ*loxP* (Liu *et al.*, 2003; the *loxP* site was deleted with *BstBI* and *BamHI* and two annealed oligos were ligated into the same sites: "*loxP* deletion sense oligo" and "*loxP* deletion antisense oligo", see Material and Methods) into pBKS- – *Kozak-iCre*-Intron-pA 3' of the artificial intron using *BamHI* and *EcoRI*, resulting in pBKS- – *Kozak–iCre*–Intron-pA–*FRT*neo*FRT*.

Subsequently, the knock-in cassette was released using *NotI* and *EcoRI* and introduced into the same sites of pBKS- *Sox17* Ex1-HR. The cassette was then cut out of the resulting plasmid pBKS- *Sox17* Ex1-HR- *Kozak-iCre*-Intron-pA-*FRT*neo*FRT* using *NsiI* and *KpnI* and introduced into pL254-*Sox17* via bacterial homologous recombination in EL350 bacteria, resulting in the final targeting construct (pL254 *Sox17-iCre*, see figure 13), which was confirmed by sequencing.

6.2.16. Detailed description of the generation of the targeting construct for *Foxa2*

Generation of the knock-in construct was carried out as follows. 5' and 3' homology regions for the *Foxa2* gene were amplified by PCR (fwd 5' HR *NotI*, rev 5' HR *HindIII-NdeI*, fwd 3' HR *HindIII*, rev 3' HR *SpeI*) using a 129Sv BAC (RPCI22-254-G2; Osoegawa *et al.*, 2000) as a template and subcloned into 5'-*NotI* and 3'-*SpeI* sites of the pL253 vector (Lee *et al.*, 2001; Liu *et al.*, 2003). Using gap repair a 10 kb genomic fragment was retrieved from the same BAC via homologous recombination in EL350 bacteria as described previously (see Material and Methods; Lee *et al.*, 2001; Liu *et al.*, 2003), resulting in pL253 *Foxa2*. *HindIII* and *NdeI* sites in between the homology regions were used as single cutters for linearization prior to gap repair.

For cloning of the knock-in cassette in pBluescript KS- (pBKS-) 5' and 3' homology regions, hereafter referred to as exon 1A and 1B respectively, for the knock-in into exon 1 of *Foxa2* were generated by PCR (EP005 *SacI-NsiI Foxa2* Exon 1A sense, EP002 *NotI Foxa2* Exon 1A reverse; EP007 *EcoRI Foxa2* Exon 1B sense, EP008 *HindIII Foxa2* Exon 1B reverse) using the above mentioned BAC as a template and subcloned into pBKS- using the introduced restriction sites (underlined), resulting in pBKS- – Exon 1 – HR. The *iCre* sequence as well as the artificial intron and the PGK promoter driven neomycin resistance gene flanked by *FRT* sites were cloned as described above (see design and generation of the *Sox17* targeting vector), resulting in pBKS- *Kozak-iCre*-Intron-pA-*FRT*neo*FRT*. Subsequently, the knock-in cassette was released using *NotI* and *EcoRI* and introduced into the same sites of pBKS- *Foxa2* Ex1-HR. The cassette was then cut out of the resulting plasmid pBKS- – *Foxa2* Ex1-HR–*Kozak-iCre*-Intron-pA-*FRT*neo*FRT* using the introduced *NsiI* site (see above) and the Bluescript-own *KpnI* site and introduced into pL253-*Foxa2* via bacterial homologous recombination in EL350 bacteria, resulting in the final targeting construct (pL253 *Foxa2-iCre*, see figure 17), which was confirmed by sequencing.

6.2.17. Detailed description of the generation of the vector for miRNA expression

The expression vector for miRNA should meet two purposes: resistance for selection in ES cells and expression of a fluorescent marker (blue) for easy determination of expression levels in different transgenic clones.
To accomplish this, the GFP from pUI4-GFP-SIBR was released using *NcoI* and *PstI*. The vector was first cut with *NcoI*, blunted with DNA polymerase I (large Klenow fragment) and afterwards cut with *PstI* to obtain one blunt end and one *PstI* 5'-overhang. The insert, H2B-CFP – IRES-Puro, was cut out of pCAGGS H2B-CFP IRES-Puro (generated by Dr. Ingo Burtscher) using *NotI* and *PstI* while the *NotI* 5' overhang was filled up the same way (see vector digest) in between the two restriction steps.

6.2.18. Generation of miRNA overexpression vectors

The oligos representing miRNAs were aligned and phosphorylated, ligated into the dephosphorylated vector using the *BglII* site and checked for correct orientation using an *EcoRI* digest, resulting in 616bp and approximately 690bp for the correct and the wrong orientation, respectively. The vectors were then sequenced by SequiServe using a forward primer upstream and a reverse primer downstream of the integration site of the miRNA (EP604, EP 605; see figure 38 and table 5).

6.3. Methods in protein biochemistry

6.3.1. Extraction of proteins

The extraction of proteins requires to work fast and at low temperatures to avoid denaturation of the proteins and the destruction of proteins by proteases. Different methods are used for the isolation of proteins of different cell compartments (cytoplasmic, nuclear fractions etc.). One takes advantage of the different compositions of the membranes and the osmotically active substances of the individual cell compartments. Protease inhibitors and working on ice or at 4°C should make sure that the proteins are kept intact.

a) Protein extraction: whole cell lysat

Cells are washed with ice-cold PBS. The PBS is removed properly. 50-100µl lysis buffer supplemented with protease inhibitor cocktail is used for the lyses of cells in one 6-well. Cells are realeased using a cell scraper and transferred to a 1.5ml reaction tube (Eppendorf) with a pipette.
For complete lysis cells are now rocked at 4°C in a shaker (Thermomixer comfort, Eppendorf) for 10min. After this incubation step the lysatec cells are centrifuged at 14.000rpm (centrifuge: 5417R, Eppendorf) and 4°C for 10 min to precipitate insoluble constituents like nuclei and parts of the membrane. The supernatant contains the proteins. It is transferred to a new pre-cooled tube and stored on –80°C or used immediately. 1µl is kept seperately for the determination of the protein concentration with BCA (see 6.2.3.).

b) Protein extraction: nucleic proteins according to the CelLytic™ NuClear™ Extraction Kit (Sigma NXTRACT)

I. Extraction from tissue (Embryo till E9.5)

0.1M DTT from a stock solution of 1M (1M DTT diluted 1:10) and 1x lysis buffer (isotonic, for more compact, adult tissue the hypotonic lysis buffer should be used) from a 10x stock solution was prepared. 10µl 0.1M DTT and 10µl Protease Inhibitor cocktail (PI) was added to 1000µl 1x lysis buffer and 1.5µl 0.1M DTT and 1.5µl PI was added to 147µl extraction buffer.
The tissue was washed twice with PBS. Afterwards it was carefully resuspended in 1000µl 1x lysis buffer including DTT and PI (volume lysis buffer = 5x volume tissue in µl: ~200µl tissue > ~1000µl lysis buffer). The tissue was homogenized with a needle until 90% of the cells were broken. The the cell suspension was centrifuged at 10.000-11.000xg for 20min. The supernatant that comprises the cytoplasmatic fragment was separated and stored at -80°C.
The pellet was resuspended in 140µl in the prepared extraction buffer and homogenized (volume extraction buffer = 2/3 volume tissue in µl: ~200µl tissue > ~140µl extraction buffer) and then incubated at 4°C for 30min while carefully shaking. Then the solution was centrifuged at 20.000-21.000xg and 4°C for 5min. The supernatant containing the nuclear protein fraction was transferred to a new tube and stored at -80°C.

II. Extraction from cells with detergent (ES cells, differentiated)

For preparations 0.1M DTT from a stock solution of 1M (1M DTT diluted 1:10) and 1x lysis buffer (isotonic, for more compact, adult tissue the hypotonic lysis buffer should be used) from a 10x stock solution was prepared. 10µl 0.1M DTT and 10µl Protease Inhibitor cocktail (PI) was added to 1000µl 1x lysis buffer and 1.5µl 0.1M DTT and 1.5µl PI was added to 147µl extraction buffer.
Cells were trypsinized and pelleted as usually. The supernatant was discarded completely. Then, the cells were carefully resuspended in 1000µl 1x lysis buffer including DTT and PI (volume lysis buffer = 5x volume tissue in µl: ~200µl cells > ~1000µl lysis buffer). The suspension was transferred to an Eppendorf tube and incubated on ice form maximal 15min. Cell lysis was controlled microscopically.
Then 10% igepal CA-630 was added (endconcentration: 0.6% > 6µl 10% igepal per 100µl cell suspension) and votexed for 10s to mix. Immediately after vortexing the suspension was centrifuged at 10.000-11.000xg und 4°C for 30min. The supernatant was transferred to a new tube and stored at -80°C (cytoplasmatic fraction). The pellet of cell nuclei was well-resuspended in 140µl prepared extraction buffer and homogenized (volume extraction buffer = 2/3 volume cells in µl: ~200µl cells > ~140µl extraction buffer). For complete lysis the suspension was incubated shaking at 4°C for 15-30min (middle to high speed). After the incubation step the suspension was centrifuged at 20.000-21.000xg and 4°C for 5min. The supernatant contains the nuclear proteins and was transferred to a new tube and stored on -80°C.

6.3.2. Determination of protein concentrations

a) Determination of protein concentrations using BCA

Proteins reduce alkaline Cu(II) to Cu(I) in a concentration dependent manner. Bicinchoninic acid (BCA) is a highly specific chromogenic reagent for Cu(I) because it forms in a violet complex with Cu(I) that has a maximum of absorption at 562nm. The absorption of this complex is directly proportional to the protein concentration, so that a quantitative photometric measurement of the protein solutions compared to a standard row with known concentration is possible.

As a standard row the following dilutions of BSA were prepared in lysis buffer: 25µg/ml, 125µg/ml, 250µg/ml and 500µg/ml (stored at -20°C). The protein samples were diluted with lysis buffer 1:50. 50µl null control (only lysis buffer), standard row and samples were mixed 1:20 with BCA solution (+1ml). The mixtures were incubated at 50°C for 15-30min till they became violet. Then they were cooled down to room temperature and measured in an Eppendorf photometer, starting with the null control and the standard dilutions. Then the dilution factor for the samples (1 in 49) was entered and the absorptions and concentrations of the samples were determined.

b) Determination of protein concentrations using Bradford assay

The Bradford assay for determination of protein concentrations is based on the absorbance shift of the Coomassie dye when it binds to protein (from red to blue > 595nm).
Protein samples were measured at OD_{600} relatively to one another. Therefore 1ml Bradford reagent was mixed with 1-5µl protein sample (addition of sample was dependent on its concentration: colour should change within 5min). The mixture was kept in the dark for 5min and measured against a blank value (lysis buffer without protein). The relative concentrations were calculated according to the directly proportional absorbance measured.

6.3.3. Western Blot

In Western Blots proteins are separated by their size via electrophoreses on a denaturing polyacrylamide gel first. Sodiumdodecylsulfate (SDS = negatively charged) covers the proteins and thereby gives those that are – over all – charged positively or negatively or that are neutral a negative charge according to their size. The proteins are then transferred onto a membrane using electric tension in a wet or semi-dry blot. On this membrane specific antibodies bind to the proteins that should be detected. These first antibodies are then detected by specific horseradish peroxidase (HRP) coupled secondary antibodies. The enzyme catalyzes the reaction of a substrate whose product is gloomy and can expose a film.

For saturation of unspecific binding sides of (mostly) the secondary antibody milk powder, bovine serum albumin (BSA) or serum of the animal in which the secondary antibody is made (no cross-reactivity expected) are used for blocking.

a) Denaturing SDS-polyacrylamid gelectrophorese

For 4 separating gels (10% - or see list below) the following mixture was used:

10.0ml	acrylamide/bisacrylamide-mixture (ready-to-use)
7.5ml	4x Tris/SDS buffer, pH8.8
12.5ml	H_2O
40.0µl	TEMED
300.0µl	APS

The mixture was immediately filled between two glass plates with a 10ml pipette and covered with water to exclude air and sharp straight border is achieved. After polymerization of the gel the water was decanted and completely suck off with a paper towel.

For 4 collecting gels (approximately 1cm deep underneath the pockets) the following mixture was prepared:

1.3ml	acrylamide/bisacrylamide-mixture (ready-to-use)
2.5ml	4x Tris/SDS buffer, pH6.8
6.2ml	H_2O
20.0µl	TEMED
100.0µl	APS

The mixture was immediately filled between the two glass plates with a 10ml pipette till the room was completely filled up then the comb was put in.

The 4x SDS loading buffer was freshly mixed with dithiothreitol (DTT) and mixed 1:3 with the samples. The mixture of samples and buffer were denatured at 95°C for 4min. and afterwards stored on ice. The comb was removed from the polymerized collecting gel and immediately rinsed with H_2O. Gels were the put into the gel chamber filled with 1x tris glycine running buffer. Then gel pockets were rinsed again to remove remaining gel filaments. The samples were applied into the gel pockets (20µl protein weight marker) and separated at 20mA for approximately 45min.

Acrylamide concentration (in %)	linear range of separation (kDa)
%	kDa
15.0	12 – 43
10.0	16 – 68
7.0	36 – 94
5.0	57 – 212

b) Immunoblot: Semi-dry Blot

The gels were released from the glass plates and equilibrated in KP buffer for 10min. The PVDF membrane was activated in methanol for 15s, then incubated in H_2O for 2min and afterwards in AP II buffer for 5min; nitrocellulose membrane was only wet with AP II.
The blot was built up as follows:

>anode
>1x blotting paper (thick) wet with AP I
>1x blotting paper (thin) wet with AP II
>PVDF-/nitrocellulose membrane
>gel
>1x blotting paper (thick) wet with KP
>kathode

The gel was blotted at 220mA (per small gel) for 20min. then the blot was broken up and the membrane was (optionally) put into Ponceau-S solution for 5min to confirm successful blotting. The membrane was washed or de-stained with H_2O.

c) Immunostaining

The membrane was blocked with blocking solution (1x TBST buffer + 5% milk powder (w/v) + 1g BSA (per 50m) for 2h. Then the membrane was incubated with the primary antibody in blocking solution for at least 2h at room temperature (or over night at 4°C). After incubation the membrane was washed 3x with 1x TBST for 15min. Afterwards it was incubated with the secondary antibody in blocking solution at room temperature for 1-2h. Again washing steps were carried out using 1x TBST: 3x short and 3x for 15min while shaking. For the enzymatic reaction the membrane with the coupled antibodies was wet from both sides with ECL solution 1 and 2 mixed on a glass plate immediately before. Then the membranes were wrapped in foil and exposed to a film for 1s to 10min. After exposure the film was developed.

6.3.4. Immunostainings

In immunostainings proteins can be detected using antibodies. The first antibody detects the protein, the second antibody detects the bound primary antibody. The second antibody is coupled to a fluorescent dye so that is t possible to detect it with the fluorescent microscope. It is also possible to directly label the first antibody.
First the cell were washed with PBS once and then fixed in 4% paraformaldehyde for 5min at 4°C. Afterwards they were washed three times with PBS for at least 5min for each washing step. At this point the cells were either stored at 4°C (covered with PBS) or directly used for a staining.
For staining the cells were blocked at least 1h with 1x TBS buffer + 1% serum + 0 1% Tween20 at room temperature (RT) or over night at 4°C. Then they were incubated with the first antibody in blocking solution (RT, at least 2h). After the incubation with the first antibody cells were thoroughly washed with PBS + 0.1% Tween20 three times for 10min at RT. When unspecific bindings of the first antibody were removed by washing the cells were incubated with the secondary antibody in blocking solution for 1h, then they were washed again as described. Finally they were sprinkled with a few drops mounting medium and covered with a cover slip. Pictures were taken at the inverse fluorescent microscope (Zeiss, AxioVert M200).

6.4. Methods in cell biology

6.4.1. (ES) cell culture

Murine ES cells self-renew in culture. They need a murine embryonic feeder layer as support, ES cell tested FCS and supplementation of the medium with LIF to prevent them from differentiation.

Material and Methods

a) Culture of primary murine embryonic fibroblasts (NIH3T3, HEK293)

Murine embryonic fibroblasts (MEF) were splitted every three to five days 1:4 – 1:6 depending on their growth speed and density. Therefore 15cm dishes of MEFs were washed with at least 10ml PBS –$MgCl_2$ and treated with 7ml trypsin-EDTA at 37°C for 5min. The reaction was stopped by adding 7ml MEF medium. A single cell suspension was achieved by pipetting up- and down 10 times. Afterwards the cell suspension was transferred to a 15 or 50ml Falcon tube and centrifuged at 250xg for 5min. The supernatant was discarded and the cells were resuspended in a suitable volume of MEF medium and plated on 5-6 new 15cm dishes.

b) Treatment of murine embryonic fibroblasts (MEF) with mitomycin C (MMC)

Murine embryonic fibroblasts are incubated with mitomycin C (MMC) to inhibit their growth in coculture with ES cells. MMC is an inhibitor of mitosis.
To treat MEF with MMC the cells were trypsinized as described before. Then the cells of five 15cm dishes were transferred to a 50ml Falcon tube in 20ml MEF medium and treated with 200µl MMC (1mg/ml) at 37°C for 45min. Every 15min the tube was inverted to prevent cells from attaching to the plastic of the tube. After 45min of incubation the cells were pelleted by centrifugation at 250xg for 5min. The supernatant was discarded and the cells were washed twice with MEF medium to remove. Afterwards the cells were plated on cell culture plates or dishes for direct use as a feeder layer for ES cells (see 4.4.2.d)) or they were frozen for later use (see 4.4.2. g)).

c) Seeding of murine embryonic fibroblasts (MEF) for ES coculture

After treatment with mitomycin C (see 6.4.1. b)) the MEFs were seeded for ES cell culture on plates and dishes in the following density.

10cm 1.5×10^6 cells
6cm 0.5×10^6 cells

d) Thawing of ES cells

To thaw ES cells a falcon tube containing 10ml pre-warmed ES medium was prepared. A cryovial of ES cells was thawn fastly in a 37°C waterbath while carefully shaking. The cells were then transferred into the prepared falcon tube and pelleted at 250xg for 4min
Cells were then resuspended in a suitable volume of ES cell medium (e.g. 5ml for one 6cm dish) and cultured on feeder cells at 37°C and 5-7% CO_2 in a humid incubator.

e) Passaging of ES cells

After two days (approximately 48 hours) in culture ES cells need to be splitted to prevent differentiation and to expand them. They can be splitted from 1:2 to 1:30 depending on their division rate.
The medium was removed and the cells were washed with a suitable volume of PBS (-Mg^{2+}/Ca^{2+}), approximately 5ml. After removing the PBS the cells were incubated with trypsin-EDTA (1ml per 6cm dish or 3ml per 10cm dish) for 5min at 37°C, 5-7%CO_2 in a humid incubator.
The dissociation of the cells was briefly controlled under the microscope and the reaction was stopped using at least the same amount of medium compared to the trypsin solution (4ml for one 6cm dish).
To get a singel cell suspension cells were well resuspended pipetting up and down for at least 10 times and transferred to a falcon tube. The cells were centrifuged for 4min at 250xg. The medium was discarded and the cell were resuspended in fresh ES cell medium. Dilutions of cells were plated on new feeder cells layers according to the splitting ratio determined before. Cells were incubated for approximately 48h at 37°C and 5-7% CO_2 in a humid incubator.

f) Cryoconservation of ES cells

For cryoconservation in liquid nitrogen ES cells are removed from the plate as described above (see 6.4.1.e)) and centrifuged. Instead of resuspending them in ES medium afterwards they were resuspended in pre-cooled freezing medium and transferred into cryovials. These cryovials were put into freezing boxes and stored at –80°C for at least 4 hours to cool the cells down to –80°C carefully (1°C/min). After 4 hours or incubation over night the vials were transferred into liquid nitrogen ($N2_{(l)}$).

6.4.2. Homologous recombination in ES cells

Homologous recombination is the basis for targeted mutagenesis and the generation of knock-in mice. Therefore ES cells are transformed by electroporation with the linearized targeting vector, selected for insertion and transformed colonies are picked and expanded.

a) Transformation of ES cells by electroporation

For one electroporation half a 10cm dish ES cells (70-80% confluent) was used. The cells were trypsinized as described before (see 6.4.1. e)). The single cell suspension was then transferred to a Falcon tube and centrifuged at 250xg for 5min. The cells were then washed with 10ml PBS ($-Mg^{2+}/Ca^{2+}$) at room temperature and centrifuged under the same conditions. After centrifugation the cell pellet was resuspended in 1.5ml ice-cold PBS (Mg^{2+}/Ca^{2+}). 0.7ml of this cell suspension were mixed with 0.1ml vector (25µg) in FBS ($-Mg^{2+}/Ca^{2+}$) to a final volume of 0.8ml. The mixture was then transferred into a pre-cooled cuvette and the electroporation was carried out under the following conditions.

(2 pulses, stored on ice in between for 1min)
Programm: 220 V
                 500 µF
Resistance: ∞

After electroporation the cuvette was kept on ice for 5min. Then the cells were transferred into a prepared prewarmed dish with feederlayer and ES cell medium (0.4ml of the cell suspension was put on one 10cm dish, the volume of one electroporation is therefore sufficient for two 10cm dishes). The medium was exchanged daily. After 24h the selection was started with neomycin (G418) in a final concentration of 300µg/ml or puromycin in a concentration up to 2µg/ml. Clones could be picked after 6-8 days.

b) Picking of ES cell clones

For picking of ES cell clones of one construct two conical 96-well plates with 60µ PBS $-Mg^{2+}/Ca^{2+}$ per well and four normal 96-well plates, two of which were coated with gelatine (0.1%) and filled with 100µl ES selection medium and two with a feeder layer and 100µl ES selection medium (see 6.4.1. c)).
The medium of one 10cm dish with ES cell clones was removed and 10ml PBS $-Mg^{2+}/Ca^{2+}$ were added. Under the stereo microscope clones that looked compact and round were picked with a 100µl pipette set on 20µl. The clones were detached by tapping with the tip of the pipette and then sucked and transferred into one well of the prepared 96-well dishes filled with PBS $-Mg^{2+}/Ca^{2+}$. It was taken care not to mix clones by detached cells. After one 96-well plate was filled 30µl trypsin-EDTA were added per well (total volume 110µl) and the cells incubated at 37°C for 15min. After incubation the cells were pipetted 10 times up and down with a multi-channel pipette to achieve a single cell suspension. 50µl were then transferred to each of the two prepared 96-well plates (one with feeder layer which is the master plate for freezing, one with gelatine-coating which is the template for DNA preparation).

c) Expansion of ES cell clones

The cells on 96-well plates were incubated at 37°C and 6% CO_2 in a humid incubator. After 2-4 days when the cells are dense enough the master plate was frozen. When the medium of the DNA plate turned yellow in one day DNA was prepared (see 6.2.1.1. b)).

d) Cryoconservation of ES cell clones in 96-well-plates

For the cryoconservation of ES cell clones in 96-well plates 2x freezing medium (10ml per 96-well plate) was prepared and cooled down to 4°C.

<u>2x ES freezing medium:</u>
4ml   ES-Zellmedium
4ml   FCS
<u>2ml   DMSO</u>
10ml

The medium was removed and the cells were washed once with 200µl PBS $-Mg^{2+}/Ca^{2+}$. Then 40µl trypsin-EDTA per well were added and cells were incubated at 37°C for 5min. With 60µl cold medium the reation was stopped and the cells were resuspended by pipetting up and down 10 times (as fast and cold as possible). Per well 100µl 2x freezing medium were added as fast as possible. The plates were closed with parafilm, put in napkins and then in a box that was stuffed with napkins. The box was stored at $-80°C$ for 6-8 weeks.

Material and Methods

I. Generation of $Sox17^{iCre/+}$ cells
The targeting vector was linearised with *DraIII* (25µg) and electroporated into TBV2 cells. Homologous recombination was confirmed by Southern Blot.

II. Generation of $Foxa2^{iCre/+}$ cells
The targeting vector was linearised with *DraIII* (25µg) and electroporated into TBV2 cells. Homologous recombination was confirmed by Southern Blot.

III. Generation of miRNA-transgenic ES cell clones
Verified miRNA vectors were linearized with *ScaI* and electroporated into ES cells carrying the knock-in for *Sox17* Cherry fusion (clone No. H21B/D12B8, IDG3.2). Cells were selected with 1µg/ml puromycin and 48 clones with different expression levels in terms of CFP fluorescence for each miRNA construct were expanded (see figure 40).

6.4.3. *In vitro* differentiation of ES cells in *Foxa2*- and *Sox17*-positive progenitor cells

For the in vitro differentiation (see Results, 4.5.1.) fibroblasts (NIH3T3) overexpressing Wnt3a were seeded in a density of $3x10^5$ cells per 6-well in feeder medium.
The same day (2h later) ES cells were seeded in SFO3 differentiation medium in a density of $6,0x10^5$ cells per 6-well. Therefore cells were trypsinized as described and cells were put twice on new cell culture dishes after centrifugation and removal of trypsin in new ES cell medium for preplating (removal of feeder cells). Afterwards the ES cells were counted using a Neubauer counting chamber. The cells were incubated at 37°C and 5% CO_2. The medium was exchanged every two days. After 5-6 days 90-100% *Foxa2/Sox17*-positive cells can be expected.

6.4.4. Ca/Phosphate transfections

In Ca/Phosphate transfections cells take up DNA aggregates that form in HBS buffer when Ca^{2+} and DNA is added. For transfections a mix of 40µg DNA and 244µmol $CaCl_2$ (e.g. 122µl 2M $CaCl_2$) was prepared and filled up to 1ml with dist. H_2O. 1ml 2x HBS buffer was tranferred to a round bottom tube and the DNA mix was carefully added dropwise while while bubbling with a pipette. Afterwards the mixture was incubated at room temperature (RT) for 10-20min. 1ml of the precipitates were dropped onto the cells and spread by circulating the dish. Finally the cells were incubated at 37°C and 5% CO_2 for 8-16h, then the medium was replaced. Expression of the transgene could be observed after 24-48h.

6.4.5. Fluorescent actived cell sorting: FACS

FACS analysis allows fort he detection, the discrimination and the sorting of cells in a cell mixture according to their size, their complexity or marker expression (detected with fluorescence/ fluorescent antibodies). One takes the advantage of the fact that cells of different size and granularity differentially scatter the light turned on them. The cells that should be analysed are given into a salt solution and hydrodynamically focused; that means, the cells are routed alongside a laser in a thin liquid stream. The laser light is scattered by the cells and the subsequent defelction is detected. The forward scatter includes information about the cell size while the side scatter is a parameter for the complexity or granularity. Using fluorescent signals single cell types can be detected and distinguished. The information are translated to signals by an analog digital converter that are analysed by a software and displayed as a frequency distribution.
Cells were trypsinized as described, resuspended in FACS buffer and filtered to removed cell clumps. Out of one cell sample at least 50.000 events (cells) were measured.

6.5. Methods in embryology

6.5.1. Mouse husbandry and –matings

Mice were kept in a day-night cycle from 6am-6pm. For determination of embryonic stages female mice were checked for vaginal plug when breeding. Noon of the day of vaginal plug was considered as embryonic day 0.5 (E0.5).

6.5.2. Genotyping of mice and embryos using PCR

a) Genotyping of $Sox17^{Cre/+}$ mice

Mice were genotyped by PCR analysis on tail tip genomic DNA (Laird et al., 1991; see Material and Methods). Genotyping of the $Sox17^{Cre\Delta neo/+}$ (4 generations backcrossed to C57Bl/6) was performed using a forward primer in the artificial intron (EP418; see figure 13 and table 5) and another one in the 5′ upstream region of the Sox17 locus (EP510; see figure 13 and table 5) and in the genomic intron 1 (EP419; see figure 13 and table 5) yielding products of 314bp and 589bp for the targeted and wild type alleles, respectively. $Sox17^{Cre/+}$ and $Sox17^{Cre\Delta neo/+}$ were genotypically distinguished using the same forward primer in the artificial intron (EP418; see figure 13 and table 5) and two reverse primers: one in the neomycin selection cassette (EP420; see figure 13 and table 5) and EP419 (see above) resulting in products with 444bp ($Sox17^{Cre}$) and 314bp ($Sox17^{Cre\Delta neo}$).

The PCR mixture and porgramm were carried out as follows.

2.00µl 10x Taq buffer (Fermentas GmbH)	95°C 5min
2.00µl 25mM MgCl$_2$	
2.00µl 10mM dNTPs	95°C 30s ⎫
1.00µl 10µM Primer 418	56°C 45s ⎬ 35x
1.00µl 10µM Primer 419	72°C 1min ⎭
1.00µl 10µM Primer 510	
0.25µl Taq (Fermentas GmbH)	72°C 10min
1.00µl genomic DNA	4°C ∞
9.75µl dist. H$_2$O	
20.00µl	

b) Genotyping of $Foxa2^{Cre/+}$ mice

For Genotyping of the $Foxa2^{Cre/+}$ mice PCR was performed on tail tip genomic DNA (for preparation see Material and Methods; Laird et al., 1991). $Foxa2^{Cre\Delta neo/+}$ (5 generations backcrossed onto C57Bl/6 background) were genotyped using a forward primer in the artificial intron (see figure 17 and table 5; EP418), another forward primer in the 5′ upstream region of the Foxa2 locus (see figure 17 and table 5; EP511) and a reverse primer in the genomic intron 1 (see figure 17 and table 5; EP421) yielding products of 415bp and 636bp for the targeted and wild type alleles, respectively. $Foxa2^{Cre/+}$ and $Foxa2^{Cre\Delta neo/+}$ were genotypically distinguished using the same forward primer in the artificial intron (see above; EP418) and two reverse primers: one in the neomycin selection cassette (see figure 17 and table 5; EP420) and EP421 (see above) resulting in products with 494bp ($Foxa2^{Cre}$) and 415bp ($Foxa2^{Cre\Delta neo}$).

The PCR mixture and porgramm were carried out as follows.

2.00µl 10x Taq buffer (Fermentas GmbH)	95°C 5min
2.00µl 25mM MgCl$_2$	
2.00µl 10mM dNTPs	95°C 30s ⎫
1.00µl 10µM Primer 418	56°C 45s ⎬ 35x
1.00µl 10µM Primer 421	72°C 1min ⎭
1.00µl 10µM Primer 511	
0.25µl Taq (Fermentas GmbH)	72°C 10min
1.00µl genomic DNA	4°C ∞
9.75µl dist. H$_2$O	
20.00µl	

c) Genotyping of R26R mice

Genotyping of R26R Cre reporter strain (background C57Bl/6) was performed by PCR as described (Soriano, 1999).

d) Genotyping of β-catenin$^{flox/+}$ or β-catenin$^{floxdel/+}$ mice

Genotyping of β-catenin$^{flox/+}$ or β-catenin$^{floxdel/+}$ mice (background C57Bl/6) was performed by PCR as described (Brault et al., 2001).

e) Genotyping of *Flp-e* mice

Genotyping of *Flp-e* mice (background C57Bl/6) was performed by PCR as described (Dymecki, 1996).

6.5.3. Isolation of embryos and organs

Dissections of embryos and organs were carried out according to Nagy and Behringer („Manipulating the mouse embryo: a laboratory manual"). Embryos were staged according to Downs and Davies (1993).

6.5.4. X-gal (5-bromo-4-chloro-3-indolyl β-D-galactoside) staining

Dissected embryos and organs were washed in PBS+Mg^{2+}/Ca^{2+}. Then they were fixed in fixation buffer for 15-60min depending on their developmental stage (size):

- a) E12.5 or earlier: 15-40min (RT, while rolling)
- b) E13.5 or later: 30-60min (RT, while rolling)

Afterwards samples were washed three times in PBS +0.02% NP40 and then placed in X-gal staining solution at 37°C over night. The next day embryos were washed again in PBS +0.02% NP40 three times and then fixed in 4% paraformaldehyde (PFA) at 4°C for 1h. Samples were stored at 4°C in 4% PFA. For documentation organs were rinsed in PBS and photographed on a Zeiss Lumar.V12 stereo using an AxioCam MRc5 camera.

6.5.5. Clearing of embryos and organs

To clear embryos and organs they were dehydrated using an ascending methanol row (25%, 50%, 75%, 100%; 10min each) and then put into benzyl alcohol/benzyl benzoate (BABB, 2:19). They were incubated in this mixture until they appeared glassy and stored in that solution for taking pictures (see X-gal staining above). Afterwards the methanol row was carried out downwards for rehydration.

4.6. Methods in histology

4.6.1. Paraffin sections

a) Embedding of embryos and organs for paraffin sections

Comment: Small embryos or small tissue pieces can be put in agarose for orientation. Before dehydration starts the tissue is embedded into 0.8% agarose in PBS and cut out as a cube. The rest of the procedure stays the same.

After whole-mount b-galactosidase staining, embryos were dehydrated with methanol (25%, 50%, 75%, 2x 100% at least 10min each). Dehydration is necessary to allow for the complete penetration of the tissue by the hydrophobic paraffin which is necessary for easy cutting. Then the tissue was put into 100% xylene till it appeared glassy. Afterwards it was incubated in paraffin at 63°C over night. The next day the tissue was put into fresh 63°C paraffin and was again incubated over night.
The following day the tissue samples were put in embedding moulds, orientated and covered with liquid paraffin. Now the paraffin was cooled down and after solidification the mould was removed. The paraffin block was fixed to an embedding grid by partial melting of the paraffin block.

b) Sectioning of paraffin blocks

The compact paraffin block was fixed to the microtom and orientated. Immediately after cutting the sections were put into 37°C water for spreading. Afterwards they were put on glass slides with a brush and dried at 37-40°C over night.

c) Histological staining of paraffin sections using Nuclear Fast Red (NFR)

Paraffin sections on glass slides were dewaxed twice for 15min in xylene. Then they were rehydrated with a downward alcohol row (100%, 90%, 80%, 70%, 1min each) and put into H_2O in the end. Then the slides were dipped into Nuclear Fast Red for 1min and thoroughly washed with dist. H_2O.
Then the sections were dehydrated with in an afferent alcohol row (70-100%, see above). Finally the slides were incubated twice in xylene for 15-30min and once in Rotihistol for 15-30min. After incubation in Rotihistol slides were put on a paper towel, sprinkled with a few drops mounting medium and covered with a cover slip.

7. References

Abremski K, Hoess R. 1984. Bacteriophage P1 site-specific recombination. Purification and properties of the Cre recombinase protein. J Bio Chem 259:1509-1514.

Adams MD, Celniker SE, Holt RA, Evans CA, Gocayne JD, Amanatides PG, Scherer SE, Li PW, Hoskins RA, Galle RF, George RA, Lewis SE, Richards S, Ashburner M, Henderson SN, Sutton GG, Wortman JR, Yandell MD, Zhang Q, Chen LX, Brandon RC, Rogers YH, Blazej RG, Champe M, Pfeiffer BD, Wan KH, Doyle C, Baxter EG, Helt G, Nelson CR, Gabor GL, Abril JF, Agbayani A, An HJ, Andrews-Pfannkoch C, Baldwin D, Ballew RM, Basu A, Baxendale J, Bayraktaroglu L, Beasley EM, Beeson KY, Benos PV, Berman BP, Bhandari D, Bolshakov S, Borkova D, Botchan MR, Bouck J, Brokstein P, Brottier P, Burtis KC, Busam DA, Butler H, Cadieu E, Center A, Chandra I, Cherry JM, Cawley S, Dahlke C, Davenport LB, Davies P, de Pablos B, Delcher A, Deng Z, Mays AD, Dew I, Dietz SM, Dodson K, Doup LE, Downes M, Dugan-Rocha S, Dunkov BC, Dunn P, Durbin KJ, Evangelista CC, Ferraz C, Ferriera S, Fleischmann W, Fosler C, Gabrielian AE, Garg NS, Gelbart WM, Glasser K, Glodek A, Gong F, Gorrell JH, Gu Z, Guan P, Harris M, Harris NL, Harvey D, Heiman TJ, Hernandez JR, Houck J, Hostin D, Houston KA, Howland TJ, Wei MH, Ibegwam C, Jalali M, Kalush F, Karpen GH, Ke Z, Kennison JA, Ketchum KA, Kimmel BE, Kodira CD, Kraft C, Kravitz S, Kulp D, Lai Z, Lasko P, Lei Y, Levitsky AA, Li J, Li Z, Liang Y, Lin X, Liu X, Mattei B, McIntosh TC, McLeod MP, McPherson D, Merkulov G, Milshina NV, Mobarry C, Morris J, Moshrefi A, Mount SM, Moy M, Murphy B, Murphy L, Muzny DM, Nelson DL, Nelson DR, Nelson KA, Nixon K, Nusskern DR, Pacleb JM, Palazzolo M, Pittman GS, Pan S, Pollard J, Puri V, Reese MG, Reinert K, Remington K, Saunders RD, Scheeler F, Shen H, Shue BC, Siden-Kiamos I, Simpson M, Skupski MP, Smith T, Spier E, Spradling AC, Stapleton M, Strong R, Sun E, Svirskas R, Tector C, Turner R, Venter E, Wang AH, Wang X, Wang ZY, Wassarman DA, Weinstock GM, Weissenbach J, Williams SM, Woodage T, Worley KC, Wu D, Yang S, Yao QA, Ye J, Yeh RF, Zaveri JS, Zhan M, Zhang G, Zhao Q, Zheng L, Zheng XH, Zhong FN, Zhong W, Zhou X, Zhu S, Zhu X, Smith HO, Gibbs RA, Myers EW, Rubin GM, Venter JC. 2000. The genome sequence of Drosophila melanogaster. Science 287:2185-2195.

Agarwal S, Holton KL, Lanza R. 2008. Efficient differentiation of functional hepatocytes from human embryonic stem cells. Stem Cells 26:1117-1127.

Alexander J, Stainier DY. 1999. A molecular pathway leading to endoderm formation in zebrafish. Curr Biol 9:1147-1157.

Alvarez-Buylla A, Seri B, Doetsch F. 2002. Identification of neural stem cells in the adult vertebrate brain. Brain Res Bull 57:751-758.

Ambros V, Bartel B, Bartel DP, Burge CB, Carrington JC, Chen X, Dreyfuss G, Eddy SR, Griffiths-Jones S, Marshall M, Matzke M, Ruvkun G, Tuschl T. 2003. A uniform system for microRNA annotation. RNA 9:277-279.

Anderson WJ, Zhou Q, Alcalde V, Kaneko OF, Blank LJ, Sherwood RI, Guseh JS, Rajagopal J, Melton DA. 2008. Genetic targeting of the endoderm with claudin-6CreER. Dev Dyn 237:504-512.

Ang SL, Conlon RA, Jin O, Rossant J. 1994. Positive and negative signals from mesoderm regulate the expression of mouse Otx2 in ectoderm explants. Development 120:2979-2989.

Ang SL, Constam DB. 2004. A gene network establishing polarity in the early mouse embryo. Semin Cell Dev Biol 15:555-561.

Ang SL, Rossant J. 1994. HNF-3 beta is essential for node and notochord formation in mouse development. Cell 78:561-574.

Ang SL, Wierda A, Wong D, Stevens KA, Cascio S, Rossant J, Zaret KS. 1993. The formation and maintenance of the definitive endoderm lineage in the mouse: involvement of HNF3/forkhead proteins. Development 119:1301-1315.

Aoi T, Yae K, Nakagawa M, Ichisaka T, Okita K, Takahashi K, Chiba T, Yamanaka S. 2008. Generation of pluripotent stem cells from adult mouse liver and stomach cells. Science 321:699-702.

Avilion AA, Nicolis SK, Pevny LH, Perez L, Vivian N, Lovell-Badge R. 2003. Multipotent cell lineages in early mouse development depend on SOX2 function. Genes Dev 17:126-140.

Bagga S, Bracht J, Hunter S, Massirer K, Holtz J, Eachus R, Pasquinelli AE. 2005. Regulation by let-7 and lin-4 miRNAs results in target mRNA degradation. Cell 122:553-563.

Baron MH. 2003. Embryonic origins of mammalian hematopoiesis. Exp Hematol 31:1160-1169.

Bartel DP. 2004. MicroRNAs: genomics, biogenesis, mechanism, and function. Cell 116:281-297.

Beachy PA, Karhadkar SS, Berman DM. 2004. Tissue repair and stem cell renewal in carcinogenesis. Nature 432:324-331.

Beck F, Erler T, Russell A, James R. 1995. Expression of Cdx-2 in the mouse embryo and placenta: possible role in patterning of the extra-embryonic membranes. Dev Dyn 204 219-227.

References

Beddington RS. 1994. Induction of a second neural axis by the mouse node. Development 120:613-620.
Beddington RS, Robertson EJ. 1998. Anterior patterning in mouse. Trends Genet 14:277-284.
Beddington RS, Robertson EJ. 1999. Axis development and early asymmetry in mammals. Cell 96:195-209.
Bedell MA, Jenkins NA, Copeland NG. 1997a. Mouse models of human disease. Part I: techniques and resources for genetic analysis in mice. Genes Dev 11:1-10.
Bedell MA, Largaespada DA, Jenkins NA, Copeland NG. 1997b. Mouse models of human disease. Part II: recent progress and future directions. Genes Dev 11:11-43.
Belo JA, Bouwmeester T, Leyns L, Kertesz N, Gallo M, Follettie M, De Robertis EM. 1997. Cerberus-like is a secreted factor with neutralizing activity expressed in the anterior primitive endoderm of the mouse gastrula. Mech Dev 68:45-57.
Bernstein E, Kim SY, Carmell MA, Murchison EP, Alcorn H, Li MZ, Mills AA, Elledge SJ, Anderson KV, Hannon GJ. 2003. Dicer is essential for mouse development. Nat Genet 35:215-217.
Bertani G. 1951. A Method for Detection of Mutations, Using Streptomycin Dependence in *Escherichia Coli*. Genetics 36:598-611.
Blum M, Gaunt SJ, Cho KW, Steinbeisser H, Blumberg B, Bittner D, De Robertis EM. 1992. Gastrulation in the mouse: the role of the homeobox gene *goosecoid*. Cell 69:1097-1106.
Bochkis IM, Rubins NE, White P, Furth EE, Friedman JR, Kaestner KH. 2008. Hepatocyte-specific ablation of Foxa2 alters bile acid homeostasis and results in endoplasmic reticulum stress. Nat Med 14:828-836.
Bohnsack MT, Czaplinski K, Gorlich D. 2004. Exportin 5 is a RanGTP-dependent dsRNA-binding protein that mediates nuclear export of pre-miRNAs. RNA 10:185-191.
Bouwmeester T, Kim S, Sasai Y, Lu B, De Robertis EM. 1996. Cerberus is a head-inducing secreted factor expressed in the anterior endoderm of Spemann's organizer. Nature 382:595-601.
Bowles J, Schepers G, Koopman P. 2000. Phylogeny of the SOX family of developmental transcription factors based on sequence and structural indicators. Dev Biol 227:239-255.
Brault V, Moore R, Kutsch S, Ishibashi M, Rowitch DH, McMahon AP, Sommer L, Boussadia O, Kemler R. 2001. Inactivation of the *beta-catenin* gene by Wnt1-Cre-mediated deletion results in dramatic brain malformation and failure of craniofacial development. Development 128:1253-1264.
Brennan J, Lu CC, Norris DP, Rodriguez TA, Beddington RS, Robertson EJ. 2001. Nodal signalling in the epiblast patterns the early mouse embryo. Nature 411:965-969.
Brennecke J, Hipfner DR, Stark A, Russell RB, Cohen SM. 2003. *bantam* encodes a developmentally regulated microRNA that controls cell proliferation and the proapoptotic gene *hid* in Drosophila. Cell 113:25-36.
Brennecke J, Stark A, Russell RB, Cohen SM. 2005. Principles of microRNA-target recognition. PLoS Biol 3:e85.
Brummelkamp TR, Bernards R, Agami R. 2002a. Stable suppression of tumorigenicity by virus-mediated RNA interference. Cancer Cell 2:243-247.
Brummelkamp TR, Bernards R, Agami R. 2002b. A system for stable expression of short interfering RNAs in mammalian cells. Science 296:550-553.
Brustle O, Jones KN, Learish RD, Karram K, Choudhary K, Wiestler OD, Duncan ID, McKay RD. 1999. Embryonic stem cell-derived glial precursors: a source of myelinating transplants. Science 285:754-756.
Calin GA, Sevignani C, Dumitru CD, Hyslop T, Noch E, Yendamuri S, Shimizu M, Rattan S, Bullrich F, Negrini M, Croce CM. 2004. Human microRNA genes are frequently located at fragile sites and genomic regions involved in cancers. Proc Natl Acad Sci U S A 101:2999-3004.
Care A, Catalucci D, Felicetti F, Bonci D, Addario A, Gallo P, Bang ML, Segnalini P, Gu Y, Dalton ND, Elia L, Latronico MV, Hoydal M, Autore C, Russo MA, Dorn GW, 2nd, Ellingsen O, Ruiz-Lozano P, Peterson KL, Croce CM, Peschle C, Condorelli G. 2007. MicroRNA-133 controls cardiac hypertrophy. Nat Med 13:613-618.
Carmeliet P, Ferreira V, Breier G, Pollefeyt S, Kieckens L, Gertsenstein M, Fahrig M, Vandenhoeck A, Harpal K, Eberhardt C, Declercq C, Pawling J, Moons L, Collen D, Risau W, Nagy A. 1996. Abnormal blood vessel development and lethality in embryos lacking a single *VEGF* allele. Nature 380:435-439.
Chaya D, Hayamizu T, Bustin M, Zaret KS. 2001. Transcription factor FoxA (HNF3) on a nucleosome at an enhancer complex in liver chromatin. J Biol Chem 276:44385-44389.
Chen CZ, Li L, Lodish HF, Bartel DP. 2004. MicroRNAs modulate hematopoietic lineage differentiation. Science 303:83-86.
Chen JF, Mandel EM, Thomson JM, Wu Q, Callis TE, Hammond SM, Conlon FL, Wang DZ. 2006. The role of microRNA-1 and microRNA-133 in skeletal muscle proliferation and differentiation. Nat Genet 38:228-233.
Cheng LC, Tavazoie M, Doetsch F. 2005. Stem cells: from epigenetics to microRNAs. Neuron 46:363-367.
Choi D, Oh HJ, Chang UJ, Koo SK, Jiang JX, Hwang SY, Lee JD, Yeoh GC, Shin HS, Lee JS, Oh B. 2002. *In vivo* differentiation of mouse embryonic stem cells into hepatocytes. Cell Transplant 11:359-368.
Choi WY, Giraldez AJ, Schier AF. 2007. Target protectors reveal dampening and balancing of Nodal agonist and antagonist by miR-430. Science 318:271-274.

Chung KH, Hart CC, Al-Bassam S, Avery A, Taylor J, Patel PD, Vojtek AB, Turner DL. 2006. Polycistronic RNA polymerase II expression vectors for RNA interference based on BIC/miR-155. Nucleic Acids Res 34:e53.

Cirillo LA, Lin FR, Cuesta I, Friedman D, Jarnik M, Zaret KS. 2002. Opening of compacted chromatin by early developmental transcription factors HNF3 (FoxA) and GATA-4. Mol Cell 9 279-289.

Clurman BE, Hayward WS. 1989. Multiple proto-oncogene activations in avian leukosis virus-induced lymphomas: evidence for stage-specific events. Mol Cell Biol 9:2657-2664.

Cockell M, Stolarczyk D, Frutiger S, Hughes GJ, Hagenbuchle O, Wellauer PK. 1995. Binding sites for hepatocyte nuclear factor 3 beta or 3 gamma and pancreas transcription factor 1 are required for efficient expression of the gene encoding pancreatic alpha-amylase. Mol Cell Biol 15:1933-1941.

Cohen-Tannoudji M, Vandormael-Fournin S, Drezen J, Mercier P, Babinet C, Morello D. 2000. lacZ sequences prevent regulated expression of housekeeping genes. Mech Dev 90:29-39.

Conlon FL, Lyons KM, Takaesu N, Barth KS, Kispert A, Herrmann B, Robertson EJ. 1994. A primary requirement for nodal in the formation and maintenance of the primitive streak in the mouse. Development 120:1919-1928.

Constantinescu S. 2003. Stemness, fusion and renewal of hematopoietic and embryonic stem cells. J Cell Mol Med 7:103-112.

Crosnier C, Stamataki D, Lewis J. 2006. Organizing cell renewal in the intestine: stem cells, signals and combinatorial control. Nat Rev Genet 7:349-359.

Cullen BR. 2004a. Derivation and function of small interfering RNAs and microRNAs. Virus Res 102:3-9.

Cullen BR. 2004b. Transcription and processing of human microRNA precursors. Mol Cell 16:861-865.

D'Amour KA, Agulnick AD, Eliazer S, Kelly OG, Kroon E, Baetge EE. 2005. Efficient differentiation of human embryonic stem cells to definitive endoderm. Nat Biotechnol 23:1534-1541.

D'Amour KA, Bang AG, Eliazer S, Kelly OG, Agulnick AD, Smart NG, Moorman MA, Kroon E, Carpenter MK, Baetge EE. 2006. Production of pancreatic hormone-expressing endocrine cells from human embryonic stem cells. Nat Biotechnol 24:1392-1401.

Dessimoz J, Bonnard C, Huelsken J, Grapin-Botton A. 2005. Pancreas-specific deletion of beta-catenin reveals Wnt-dependent and Wnt-independent functions during development. Curr Biol 15:1677-1683.

Dessimoz J, Opoka R, Kordich JJ, Grapin-Botton A, Wells JM. 2006. FGF signaling is necessary for establishing gut tube domains along the anterior-posterior axis in vivo. Mech Dev 123:42-55.

Deutsch G, Jung J, Zheng M, Lora J, Zaret KS. 2001. A bipotential precursor population for pancreas and liver within the embryonic endoderm. Development 128:871-881.

Didiano D, Hobert O. 2006. Perfect seed pairing is not a generally reliable predictor for miRNA-target interactions. Nat Struct Mol Biol 13:849-851.

Doench JG, Petersen CP, Sharp PA. 2003. siRNAs can function as miRNAs. Genes Dev 17:438-442.

Donovan PJ, de Miguel MP. 2003. Turning germ cells into stem cells. Curr Opin Genet Dev 13:463-471.

Donovan PJ, Gearhart J. 2001. The end of the beginning for pluripotent stem cells. Nature 414:92-97.

Downs KM, Davies T. 1993. Staging of gastrulating mouse embryos by morphological landmarks in the dissecting microscope. Development 118:1255-1266.

Dufort D, Schwartz L, Harpal K, Rossant J. 1998. The transcription factor HNF3beta is required in visceral endoderm for normal primitive streak morphogenesis. Development 125:3015-3025.

Dugas DV, Bartel B. 2004. MicroRNA regulation of gene expression in plants. Curr Opin Plant Biol 7:512-520.

Duncan AW, Rattis FM, DiMascio LN, Congdon KL, Pazianos G, Zhao C, Yoon K, Cook JM, Willert K, Gaiano N, Reya T. 2005. Integration of Notch and Wnt signaling in hematopoietic stem cell maintenance. Nat Immunol 6:314-322.

Durcova-Hills G, Ainscough J, McLaren A. 2001. Pluripotential stem cells derived from migrating primordial germ cells. Differentiation 68:220-226.

Dymecki SM. 1996. Flp recombinase promotes site-specific DNA recombination in embryonic stem cells and transgenic mice. Proc Natl Acad Sci U S A 93:6191-6196.

Eis PS, Tam W, Sun L, Chadburn A, Li Z, Gomez MF, Lund E, Dahlberg JE. 2005. Accumulation of miR-155 and BIC RNA in human B cell lymphomas. Proc Natl Acad Sci U S A 102:3627-3632.

Elbashir SM, Harborth J, Lendeckel W, Yalcin A, Weber K, Tuschl T 2001. Duplexes of 21-nucleotide RNAs mediate RNA interference in cultured mammalian cells. Nature 411:494-498.

Evans MJ, Kaufman MH. 1981. Establishment in culture of pluripotential cells from mouse embryos. Nature 292:154-156.

Farh KK, Grimson A, Jan C, Lewis EP, Johnston WK, Lim LP, Burge CE, Bartel DP. 2005. The widespread impact of mammalian MicroRNAs on mRNA repression and evolution. Science 310:1817-1821.

Fehling HJ, Lacaud G, Kubo A, Kennedy M, Robertson S, Keller G, Kouskoff V. 2003. Tracking mesoderm induction and its specification to the hemangioblast during embryonic stem cell differentiation. Development 130:4217-4227.

Feil R, Brocard J, Mascrez B, LeMeur M, Metzger D, Chambon P. 1996. Ligand-activated site-specific recombination in mice. Proc Natl Acad Sci U S A 93:10887-10890.

References

Feil R, Wagner J, Metzger D, Chambon P. 1997. Regulation of Cre recombinase activity by mutated estrogen receptor ligand-binding domains. Biochem Biophys Res Commun 237:752-757.

Ferrara N, Carver-Moore K, Chen H, Dowd M, Lu L, O'Shea KS, Powell-Braxton L, Hillan KJ, Moore MW. 1996. Heterozygous embryonic lethality induced by targeted inactivation of the VEGF gene. Nature 380:439-442.

Fisher, R. A. 1922. On the interpretation of χ^2 from contingency tables, and the calculation of P. Journal of the Royal Statistical Society. 85(1):87-94.

Fisher, R. A. 1954. Statistical Methods for research workers. Edinburgh (Oliver and Boyd) 12:356

Ford ES, Giles WH, Dietz WH. 2002. Prevalence of the metabolic syndrome among US adults: findings from the third National Health and Nutrition Examination Survey. JAMA 287:356-359.

Frank DU, Elliott SA, Park EJ, Hammond J, Saijoh Y, Moon AM. 2007. System for inducible expression of cre-recombinase from the *Foxa2* locus in endoderm, notochord, and floor plate. Dev Dyn 236:1085-1092.

Friedman JR, Kaestner KH. 2006. The Foxa family of transcription factors in development and metabolism. Cell Mol Life Sci 63:2317-2328.

Glinka A, Wu W, Delius H, Monaghan AP, Blumenstock C, Niehrs C. 1998. Dickkopf-1 is a member of a new family of secreted proteins and functions in head induction. Nature 391:357-362.

Gong S, Yang XW, Li C, Heintz N. 2002. Highly Efficient Modification of Bacterial Artificial Chromosomes (BACs) Using Novel Shuttle Vectors Containing the R6K Origin of Replication. Genome Res. 12:1992-1998

Gossler A, Joyner AL, Rossant J, Skarnes WC. 1989. Mouse embryonic stem cells and reporter constructs to detect developmentally regulated genes. Science 244:463-465.

Graham FL, Smiley J, Russell WC, Nairn R. 1977. Characteristics of a human cell line transformed by DNA from human adenovirus type 5. J Gen Virol 36:59-74.

Grapin-Botton A. 2005. Antero-posterior patterning of the vertebrate digestive tract: 40 years after Nicole Le Douarin's PhD thesis. Int J Dev Biol 49:335-347.

Green RP, Cohn SM, Sacchettini JC, Jackson KE, Gordon JI. 1992. The mouse *intestinal fatty acid binding protein* gene: nucleotide sequence, pattern of developmental and regional expression, and proposed structure of its protein product. DNA Cell Biol 11:31-41.

Grimm D, Streetz KL, Jopling CL, Storm TA, Pandey K, Davis CR, Marion P, Salazar F, Kay MA. 2006. Fatality in mice due to oversaturation of cellular microRNA/short hairpin RNA pathways. Nature 441:537-541.

Gu H, Zou YR, Rajewsky K. 1993. Independent control of immunoglobulin switch recombination at individual switch regions evidenced through *Cre-loxP*-mediated gene targeting. Cell 73:1155-1164.

Gualdi R, Bossard P, Zheng M, Hamada Y, Coleman JR, Zaret KS. 1996. Hepatic specification of the gut endoderm *in vitro*: cell signaling and transcriptional control. Genes Dev 10:1670-1682.

Haas J, Park EC, Seed B. 1996. Codon usage limitation in the expression of HIV-1 envelope glycoprotein. Curr Biol 6:315-324.

Haegel H, Larue L, Ohsugi M, Fedorov L, Herrenknecht K, Kemler R. 1995. Lack of beta-catenin affects mouse development at gastrulation. Development 121:3529-3537.

Hammond SM, Bernstein E, Beach D, Hannon GJ. 2000. An RNA-directed nuclease mediates post-transcriptional gene silencing in *Drosophila* cells. Nature 404:293-296.

Hammond SM, Boettcher S, Caudy AA, Kobayashi R, Hannon GJ. 2001. Argonaute2, a link between genetic and biochemical analyses of RNAi. Science 293:1146-1150.

Han M. 1997. Gut reaction to Wnt signaling in worms. Cell 90:581-584.

Hanahan, D. (1985). Techniques for transformation of *E. coli*. In DNA cloning techniques: a practical approach, D. M. Glover, Ed. (IRL Press, Oxford) 1: 109-135

Hartenstein, V., G.M. Technau, and].A. Campos-Orterga. 1985. Fate-mapping in wild type *Drosophila melanogaster*. Wilhelm Roux's Arch. Dev. Biol. 194: 181-195.

Hatfield SD, Shcherbata HR, Fischer KA, Nakahara K, Carthew RW, Ruohola-Baker H. 2005. Stem cell division is regulated by the microRNA pathway. Nature 435:974-978.

He L, Hannon GJ. 2004. MicroRNAs: small RNAs with a big role in gene regulation. Nat Rev Genet 5:522-531.

He X, Saint-Jeannet JP, Wang Y, Nathans J, Dawid I, Varmus H. 1997. A member of the Frizzled protein family mediating axis induction by Wnt-5A. Science 275:1652-1654.

Heinrich PC, Behrmann I, Muller-Newen G, Schaper F, Graeve L. 1998. Interleukin-6-type cytokine signalling through the gp130/Jak/STAT pathway. Biochem J 334 (Pt 2):297-314.

Hensen V. 1876. Beobachtungen über die Befruchtung und Entwicklung des Kaninchens und Meerschweinchens. Z Anat Entwickl Gesch 1:213–273 und 353–423.

Hirsinger E, Jouve C, Dubrulle J, Pourquie O. 2000. Somite formation and patterning. Int Rev Cytol 198:1-65.

Hoess R, Abremski K, Irwin S, Kendall M, Mack A. 1990. DNA specificity of the Cre recombinase resides in the 25 kDa carboxyl domain of the protein. J Mol Biol 216:873-882.

Hoess RH, Ziese M, Sternberg N. 1982. P1 site-specific recombination: nucleotide sequence of the recombining sites. Proc Natl Acad Sci U S A 79:3398-3402.

Hong YK, Shin JW, Detmar M. 2004. Development of the lymphatic vascular system: a mystery unravels. Dev Dyn 231:462-473.

References

Horb ME, Slack JM. 2001. Endoderm specification and differentiation in *Xenopus* embryos. Dev Biol 236:330-343.

Houbaviy HB, Murray MF, Sharp PA. 2003. Embryonic stem cell-specific MicroRNAs. Dev Cell 5:351-358.

Huelsken J, Vogel R, Brinkmann V, Erdmann B, Birchmeier C, Birchmeier W. 2000 Requirement for beta-catenin in anterior-posterior axis formation in mice. J Cell Biol 148:567-578.

Hutvagner G, Zamore PD. 2002. A microRNA in a multiple-turnover RNAi enzyme complex. Science 297:2056-2060.

Hwang HW, Mendell JT. 2006. MicroRNAs in cell proliferation, cell death, and tumorigenesis. Br J Cancer 94:776-780.

Iles SA. 1977. Mouse teratomas and embryoid bodies: their induction and differentiation. J Embryol Exp Morphol 38:63-75.

Iorio MV, Ferracin M, Liu CG, Veronese A, Spizzo R, Sabbioni S, Magri E, Pedriali M, Fabbri M, Campiglio M, Menard S, Palazzo JP, Rosenberg A, Musiani P, Volinia S, Nenci I, Calin GA, Querzoli P, Negrini M, Croce CM. 2005. MicroRNA gene expression deregulation in human breast cancer. Cancer Res 65:7065-7070.

Jazag A, Kanai F, Ijichi H, Tateishi K, Ikenoue T, Tanaka Y, Ohta M, Imamura J, Guleng B, Asaoka Y, Kawabe T, Miyagishi M, Taira K, Omata M. 2005. Single small-interfering RNA expression vector for silencing multiple transforming growth factor-beta pathway components Nucleic Acids Res 33:e131.

Jing Q, Huang S, Guth S, Zarubin T, Motoyama A, Chen J, Di Padova F, Lin SC, Gram H, Han J. 2005. Involvement of microRNA in AU-rich element-mediated mRNA instability. Cell 120:623-634.

Joglekar MV, Parekh VS, Mehta S, Bhonde RR, Hardikar AA. 2007. MicroRNA profiling of developing and regenerating pancreas reveal post-transcriptional regulation of *neurogenin3*. Dev Biol 311:603-612.

Jürgens, G. and D. Weigel. 1988. Terminal versus segmental development in the *Drosophila* embryo: The role of the homeotic gene *fork head*. Wilhelm Roux's Arch Dev. Biol. 197: 345-354.

Kaestner KH, Katz J, Liu Y, Drucker DJ, Schutz G. 1999. Inactivation of the winged helix transcription factor HNF3alpha affects glucose homeostasis and islet glucagon gene expression *in vivo*. Genes Dev 13:495-504.

Kaestner KH, Knochel W, Martinez DE. 2000. Unified nomenclature for the winged helix/forkhead transcription factors. Genes Dev 14:142-146.

Kanai Y, Kanai-Azuma M, Noce T, Saido TC, Shiroishi T, Hayashi Y, Yazaki K. 1996. Identification of two *Sox17* messenger RNA isoforms, with and without the high mobility group box region, and their differential expression in mouse spermatogenesis. J Cell Biol 133:667-681.

Kanai-Azuma M, Kanai Y, Gad JM, Tajima Y, Taya C, Kurohmaru M, Sanai Y, Yonekawa H, Yazaki K, Tam PP, Hayashi Y. 2002. Depletion of definitive gut endoderm in *Sox17*-null mutant mice. Development 129:2367-2379.

Kaneko-Ishino T, Kuroiwa Y, Miyoshi N, Kohda T, Suzuki R, Yokoyama M, Viville S, Barton SC, Ishino F, Surani MA. 1995. *Peg1/Mest* imprinted gene on chromosome 6 identified by cDNA subtraction hybridization. Nat Genet 11:52-59.

Kanellopoulou C, Muljo SA, Kung AL, Ganesan S, Drapkin R, Jenuwein T, Livingston DM, Rajewsky K. 2005. Dicer-deficient mouse embryonic stem cells are defective in differentiation and centromeric silencing. Genes Dev 19:489-501.

Kano M, Igarashi H, Saito I, Masuda M. 1998. Cre-loxP-mediated DNA flip-flop in mammalian cells leading to alternate expression of retrovirally transduced genes. Biochem Biophys Res Commun 248:806-811.

Kapsimali M, Caneparo L, Houart C, Wilson SW. 2004. Inhibition of Wnt/Axin/beta-catenin pathway activity promotes ventral CNS midline tissue to adopt hypothalamic rather than floorplate identity. Development 131:5923-5933.

Kaufman MH, Webb S. 1990. Postimplantation development of tetraploid mouse embryos produced by electrofusion. Development 110:1121-1132.

Kawasaki H, Taira K. 2003. Short hairpin type of dsRNAs that are controlled by tRNA(Val) promoter significantly induce RNAi-mediated gene silencing in the cytoplasm of human cells. Nucleic Acids Res 31:700-707.

Keller G. 2005. Embryonic stem cell differentiation: emergence of a new era in biology and medicine. Genes Dev 19:1129-1155.

Keller JR, Ortiz M, Ruscetti FW. 1995. Steel factor (c-kit ligand) promotes the survival of hematopoietic stem/progenitor cells in the absence of cell division. Blood 86:1757-1764.

Kim I, Saunders TL, Morrison SJ. 2007. Sox17 dependence distinguishes the transcriptional regulation of fetal from adult hematopoietic stem cells. Cell 130:470-483.

Kimelman D, Griffin KJ. 2000. Vertebrate mesendoderm induction and patterning. Curr Opin Genet Dev 10:350-356.

Kimura-Yoshida C, Tian E, Nakano H, Amazaki S, Shimokawa K, Rossant J, Aizawa S, Matsuo I. 2007. Crucial roles of Foxa2 in mouse anterior-posterior axis polarization via regulation of anterior visceral endoderm-specific genes. Proc Natl Acad Sci U S A 104:5919-5924.

Kimura C, Yoshinaga K, Tian E, Suzuki M, Aizawa S, Matsuo I. 2000. Visceral endoderm mediates forebrain

References

development by suppressing posteriorizing signals. Dev Biol 225:304-321.
Kinder SJ, Tsang TE, Wakamiya M, Sasaki H, Behringer RR, Nagy A, Tam PP. 2001. The organizer of the mouse gastrula is composed of a dynamic population of progenitor cells for the axial mesoderm. Development 128:3623-3634.
King T, Bland Y, Webb S, Barton S, Brown NA. 2002. Expression of *Peg1* (*Mest*) in the developing mouse heart: involvement in trabeculation. Dev Dyn 225:212-215.
Kispert A, Herrmann BG. 1994. Immunohistochemical analysis of the Brachyury protein in wild-type and mutant mouse embryos. Dev Biol 161:179-193.
Kittappa R, Chang WW, Awatramani RB, McKay RD. 2007. The *Foxa2* gene controls the birth and spontaneous degeneration of dopamine neurons in old age. PLoS Biol 5:e325.
Kleinsmith LJ, Pierce GB, Jr. 1964. Multipotentiality of Single Embryonal Carcinoma Cells. Cancer Res 24:1544-1551.
Kluiver J, Poppema S, de Jong D, Blokzijl T, Harms G, Jacobs S, Kroesen BJ, van den Berg A. 2005. *BIC* and *miR-155* are highly expressed in Hodgkin, primary mediastinal and diffuse large B cell lymphomas. J Pathol 207:243-249.
Korinek V, Barker N, Morin PJ, van Wichen D, de Weger R, Kinzler KW, Vogelstein B, Clevers H. 1997. Constitutive transcriptional activation by a beta-catenin-Tcf complex in APC-/- colon carcinoma. Science 275:1784-1787.
Korinek V, Barker N, Willert K, Molenaar M, Roose J, Wagenaar G, Markman M, Lamers W, Destree O, Clevers H. 1998. Two members of the Tcf family implicated in Wnt/beta-catenin signaling during embryogenesis in the mouse. Mol Cell Biol 18:1248-1256.
Kothary R, Clapoff S, Darling S, Perry MD, Moran LA, Rossant J. 1989. Inducible expression of an *hsp68-lacZ* hybrid gene in transgenic mice. Development 105:707-714.
Kozak M. 1997. Recognition of AUG and alternative initiator codons is augmented by G in position +4 but is not generally affected by the nucleotides in positions +5 and +6. EMBO J 16:2482-2492.
Kroon E, Martinson LA, Kadoya K, Bang AG, Kelly OG, Eliazer S, Young H, Richardson M, Smart NG, Cunningham J, Agulnick AD, D'Amour KA, Carpenter MK, Baetge EE. 2008. Pancreatic endoderm derived from human embryonic stem cells generates glucose-responsive insulin-secreting cells *in vivo*. Nat Biotechnol 26:443-452.
Kühn R, Schwenk F. 1997. Conditional Knockout Mice. Meth in Mol Biol, 209:159-185
Kumar A, Yamaguchi T, Sharma P, Kuehn MR. 2007. Transgenic mouse lines expressing Cre recombinase specifically in posterior notochord and notochord. Genesis 45:729-736.
Kwon C, Han Z, Olson EN, Srivastava D. 2005. MicroRNA1 influences cardiac differentiation in *Drosophila* and regulates Notch signaling. Proc Natl Acad Sci U S A 102:18986-18991.
Labosky PA, Barlow DP, Hogan BL. 1994. Embryonic germ cell lines and their derivation from mouse primordial germ cells. Ciba Found Symp 182:157-168; discussion 168-178.
Lagos-Quintana M, Rauhut R, Yalcin A, Meyer J, Lendeckel W, Tuschl T. 2002. Identification of tissue-specific microRNAs from mouse. Curr Biol 12:735-739.
Lai E, Prezioso VR, Tao WF, Chen WS, Darnell JE, Jr. 1991. *Hepatocyte nuclear factor 3 alpha* belongs to a gene family in mammals that is homologous to the *Drosophila* homeotic gene *fork head*. Genes Dev 5:416-427.
Lakso M, Sauer B, Mosinger B, Jr., Lee EJ, Manning RW, Yu SH, Mulder KL, Westphal H. 1992. Targeted oncogene activation by site-specific recombination in transgenic mice. Proc Natl Acad Sci U S A 89:6232-6236.
Lam KP, Rajewsky K. 1998. Rapid elimination of mature autoreactive B cells demonstrated by Cre-induced change in B cell antigen receptor specificity *in vivo*. Proc Natl Acad Sci U S A 95:13171-13175.
Lander ES, Linton LM, Birren B, Nusbaum C, Zody MC, Baldwin J, Devon K, Dewar K, Doyle M, FitzHugh W, Funke R, Gage D, Harris K, Heaford A, Howland J, Kann L, Lehoczky J, LeVine R, McEwan P, McKernan K, Meldrim J, Mesirov JP, Miranda C, Morris W, Naylor J, Raymond C, Rosetti R, Santos R, Sheridan A, Sougnez C, Stange-Thomann N, Stojanovic N, Subramanian A, Wyman D, Rogers J, Sulston J, Ainscough R, Beck S, Bentley D, Burton J, Clee C, Carter N, Coulson A, Deadman R, Deloukas P, Dunham A, Dunham I, Durbin R, French L, Grafham D, Gregory S, Hubbard T, Humphray S, Hunt A, Jones M, Lloyd C, McMurray A, Matthews L, Mercer S, Milne S, Mullikin JC, Mungall A, Plumb R, Ross M, Shownkeen R, Sims S, Waterston RH, Wilson RK, Hillier LW, McPherson JD, Marra MA, Mardis ER, Fulton LA, Chinwalla AT, Pepin KH, Gish WR, Chissoe SL, Wendl MC, Delehaunty KD, Miner TL, Delehaunty A, Kramer JB, Cook LL, Fulton RS, Johnson DL, Minx PJ, Clifton SW, Hawkins T, Branscomb E, Predki P, Richardson P, Wenning S, Slezak T, Doggett N, Cheng JF, Olsen A, Lucas S, Elkin C, Uberbacher E, Frazier M, Gibbs RA, Muzny DM, Scherer SE, Bouck JB, Sodergren EJ, Worley KC, Rives CM, Gorrell JH, Metzker ML, Naylor SL, Kucherlapati RS, Nelson DL, Weinstock GM, Sakaki Y, Fujiyama A, Hattori M, Yada T, Toyoda A, Itoh T, Kawagoe C, Watanabe H, Totoki Y, Taylor T, Weissenbach J, Heilig R, Saurin W, Artiguenave F, Brottier P, Bruls T, Pelletier E, Robert C, Wincker P, Smith DR, Doucette-Stamm L, Rubenfield M, Weinstock K, Lee HM, Dubois J, Rosenthal A, Platzer M, Nyakatura G, Taudien S, Rump A, Yang H, Yu J, Wang J, Huang G, Gu J, Hood L, Rowen L, Madan A, Qin S, Davis RW, Federspiel

References

NA, Abola AP, Proctor MJ, Myers RM, Schmutz J, Dickson M, Grimwood J, Cox DR, Olson MV, Kaul R, Raymond C, Shimizu N, Kawasaki K, Minoshima S, Evans GA, Athanasiou M, Schultz R, Roe BA, Chen F, Pan H, Ramser J, Lehrach H, Reinhardt R, McCombie WR, de la Bastide M, Dedhia N, Blocker H, Hornischer K, Nordsiek G, Agarwala R, Aravind L, Bailey JA, Bateman A, Batzoglou S, Birney E, Bork P, Brown DG, Burge CB, Cerutti L, Chen HC, Church D, Clamp M, Copley RR, Doerks T, Eddy SR, Eichler EE, Furey TS, Galagan J, Gilbert JG, Harmon C, Hayashizaki Y, Haussler D, Hermjakob H, Hokamp K, Jang W, Johnson LS, Jones TA, Kasif S, Kaspryzk A, Kennedy S, Kent WJ, Kitts P, Koonin EV, Korf I, Kup D, Lancet D, Lowe TM, McLysaght A, Mikkelsen T, Moran JV, Mulder N, Pollara VJ, Ponting CP, Schuler G, Schultz J, Slater G, Smit AF, Stupka E, Szustakowski J, Thierry-Mieg D, Thierry-Mieg J, Wagner L, Wallis J, Wheeler R, Williams A, Wolf YI, Wolfe KH, Yang SP, Yeh RF, Collins F, Guyer MS, Peterson J, Felsenfeld A, Wetterstrand KA, Patrinos A, Morgan MJ, de Jong P, Catanese JJ, Osoegawa K, Shizuya H, Choi S, Chen YJ. 2001. Initial sequencing and analysis of the human genome. Nature 409:860-921.

Landgraf P, Rusu M, Sheridan R, Sewer A, Iovino N, Aravin A, Pfeffer S, Rice A, Kamphorst AO, Landthaler M, Lin C, Socci ND, Hermida L, Fulci V, Chiaretti S, Foa R, Schliwka J, Fuchs U, Novosel A, Muller RU, Schermer B, Bissels U, Inman J, Phan Q, Chien M, Weir DB, Choksi R, De Vita G, Frezzetti D, Trompeter HI, Hornung V, Teng G, Hartmann G, Palkovits M, Di Lauro R, Wernet P, Macino G, Rogler CE, Nagle JW, Ju J, Papavasiliou FN, Benzing T, Lichter P, Tam W, Brownstein MJ, Bosio A, Borkhardt A, Russo JJ, Sander C, Zavolan M, Tuschl T. 2007. A mammalian microRNA expression atlas based on small RNA library sequencing. Cell 129:1401-1414.

Lantz KA, Vatamaniuk MZ, Brestell JE, Friedman JR, Matschinsky FM, Kaestner KH. 2004. Foxa2 regulates multiple pathways of insulin secretion. J Clin Invest 114:512-520.

Lawson KA, Meneses JJ, Pederser RA. 1991. Clonal analysis of epiblast fate during germ layer formation in the mouse embryo. Development 113:891-911.

Lee EC, Yu D, Martinez de Velasco J, Tessarollo L, Swing DA, Court DL, Jenkins NA, Copeland NG. 2001. A highly efficient *Escherichia coli*-based chromosome engineering system adapted for recombinogenic targeting and subcloning of BAC DNA. Genomics 73:56-65.

Lee CS, Friedman JR, Fulmer JT, Kaestner KH. 2005a. The initiation of liver development is dependent on Foxa transcription factors. Nature 435:944-947.

Lee CS, Sund NJ, Behr R, Herrera PL, Kaestner KH. 2005b. Foxa2 is required for the differentiation of pancreatic alpha-cells. Dev Biol 278:484-495.

Lee CS, Sund NJ, Vatamaniuk MZ, Matschinsky FM, Stoffers DA, Kaestner KH. 2002a. Foxa2 controls *Pdx1* gene expression in pancreatic beta-cells *in vivo*. Diabetes 51:2546-2551.

Lee NS, Dohjima T, Bauer G, Li H, Li MJ, Ehsani A, Salvaterra P, Rossi J. 2002b. Expression of small interfering RNAs targeted against HIV-1 rev transcripts in human cells. Nat Biotechnol 20:500-505.

Lee RC, Feinbaum RL, Ambros V. 1993. The *C. elegans* heterochronic gene *lin-4* encodes small RNAs with antisense complementarity to *lin-14*. Cell 75:843-854.

Lefebvre L, Viville S, Barton SC, Ishino F, Keverne EB, Surani MA. 1998. Abnormal maternal behaviour and growth retardation associated with loss of the imprinted gene *Mest*. Nat Genet 20:163-169.

Lefebvre V, Dumitriu B, Penzo-Mendez A, Han Y, Pallavi B. 2007. Control of cell fate and differentiation by Sry-related high-mobility-group box (Sox) transcription factors. Int J Biochem Cell Biol 39:2195-2214.

Lei EP, Silver PA. 2002. Protein and RNA export from the nucleus. Dev Cell 2:261-272.

Leuschner PJ, Ameres SL, Kueng S, Martinez J. 2006. Cleavage of the siRNA passenger strand during RISC assembly in human cells. EMBO Rep 7:314-320.

Lewis BP, Burge CB, Bartel DP. 2005. Conserved seed pairing, often flanked by adenosines, indicates that thousands of human genes are microRNA targets. Cell 120:15-20.

Lewis DL, Hagstrom JE, Loomis AG, Wolff JA, Herweijer H. 2002. Efficient delivery of siRNA for inhibition of gene expression in postnatal mice. Nat Genet 32:107-108.

Lewis SL, Tam PP. 2006. Definitive endoderm of the mouse embryo: formation, cell fates, and morphogenetic function. Dev Dyn 235:2315-2329.

Liao P, Uetzmann L, Burtscher I, Lickert H. 2009. Generation of a Mouse Line Expressing *Sox17*-driven Cre Recombinase with Specific Activity in Arteries. Genesis; 47(7):476-83.

Lickert H, Kutsch S, Kanzler B, Tamai Y, Taketo MM, Kemler R. 2002. Formation of multiple hearts in mice following deletion of *beta-catenin* in the embryonic endoderm. Dev Cell 3:171-181.

Lim LP, Lau NC, Garrett-Engele P, Grimson A, Schelter JM, Castle J, Bartel DP, Linsley PS, Johnson JM. 2005. Microarray analysis shows that some microRNAs downregulate large numbers of target mRNAs. Nature 433:769-773.

Lin R, Thompson S, Priess JR. 1995. Pop-1 encodes an HMG box protein required for the specification of a mesoderm precursor in early *C. elegans* embryos. Cell 83:599-609.

Liu J, Valencia-Sanchez MA, Hannon GJ, Parker R. 2005. MicroRNA-dependent localization of targeted mRNAs to mammalian P-bodies. Nat Cell Biol 7:719-723.

Liu P, Jenkins NA, Copeland NG. 2003. A highly efficient recombineering-based method for generating conditional knockout mutations. Genome Res 13:476-484.

References

Liu P, Wakamiya M, Shea MJ, Albrecht U, Behringer RR, Bradley A. 1999. Requirement for Wnt3 in vertebrate axis formation. Nat Genet 22:361-365.

Lowe LA, Yamada S, Kuehn MR. 2001. Genetic dissection of nodal function in patterning the mouse embryo. Development 128:1831-1843.

Lu CC, Brennan J, Robertson EJ. 2001. From fertilization to gastrulation: axis formation in the mouse embryo. Curr Opin Genet Dev 11:384-392.

Lund E, Guttinger S, Calado A, Dahlberg JE, Kutay U. 2004. Nuclear export of microRNA precursors. Science 303:95-98.

Makeyev EV, Zhang J, Carrasco MA, Maniatis T. 2007. The MicroRNA miR-124 promotes neuronal differentiation by triggering brain-specific alternative pre-mRNA splicing. Mol Cell 27:435-448.

Maquat LE. 2002. Nonsense-mediated mRNA decay. Curr Biol 12:R196-197.

Martello G, Zacchigna L, Inui M, Montagner M, Adorno M, Mamidi A, Morsut L, Soligo S, Tran U, Dupont S, Cordenonsi M, Wessely O, Piccolo S. 2007. MicroRNA control of Nodal signalling. Nature 449:183-188.

Martin GR. 1981. Isolation of a pluripotent cell line from early mouse embryos cultured in medium conditioned by teratocarcinoma stem cells. Proc Natl Acad Sci U S A 78:7634-7638.

Martinez J, Patkaniowska A, Urlaub H, Luhrmann R, Tuschl T. 2002. Single-stranded antisense siRNAs guide target RNA cleavage in RNAi. Cell 110:563-574.

Mason RJ, Williams MC, Moses HL, Mohla S, Berberich MA. 1997. Stem cells in lung development, disease, and therapy. Am J Respir Cell Mol Biol 16:355-363.

Matranga C, Tomari Y, Shin C, Bartel DP, Zamore PD. 2005. Passenger-strand cleavage facilitates assembly of siRNA into Ago2-containing RNAi enzyme complexes. Cell 123:607-620.

Matsui T, Kanai-Azuma M, Hara K, Matoba S, Hiramatsu R, Kawakami H, Kurohmaru M, Koopman P, Kanai Y. 2006. Redundant roles of Sox17 and Sox18 in postnatal angiogenesis in mice. J Cell Sci 119:3513-3526.

Matsui Y, Zsebo K, Hogan BL. 1992. Derivation of pluripotential embryonic stem cells from murine primordial germ cells in culture. Cell 70:841-847.

Matsuura R, Kogo H, Ogaeri T, Miwa J, Kuwahara M, Kanai Y, Nakagawa T, Kuroiwa A, Fujimoto T, Torihashi S. 2006. Crucial transcription factors in endoderm and embryonic gut development are expressed in gut-like structures from mouse ES cells. Stem Cells 24:624-630.

McCaffrey AP, Meuse L, Pham TT, Conklin DS, Hannon GJ, Kay MA. 2002. RNA interference in adult mice. Nature 418:38-39.

McLin VA, Rankin SA, Zorn AM. 2007. Repression of Wnt/beta-catenin signaling in the anterior endoderm is essential for liver and pancreas development. Development 134:2207-2217.

McMahon AP, Moon RT. 1989. Ectopic expression of the proto-oncogene int-1 in Xenopus embryos leads to duplication of the embryonic axis. Cell 58:1075-1084.

McManus MT, Haines BB, Dillon CP, Whitehurst CE, van Parijs L, Chen J, Sharp PA. 2002a. Small interfering RNA-mediated gene silencing in T lymphocytes. J Immunol 169:5754-5760.

McManus MT, Petersen CP, Haines BB, Chen J, Sharp PA. 2002b. Gene silencing using micro-RNA designed hairpins. RNA 8:842-850.

Metzger D, Clifford J, Chiba H, Chambon P. 1995. Conditional site-specific recombination in mammalian cells using a ligand-dependent chimeric Cre recombinase. Proc Natl Acad Sci U S A 92:6991-6995.

Mitsui K, Tokuzawa Y, Itoh H, Segawa K, Murakami M, Takahashi K, Maruyama M, Maeda M, Yamanaka S. 2003. The homeoprotein Nanog is required for maintenance of pluripotency in mouse epiblast and ES cells. Cell 113:631-642.

Miyagishi M, Taira K. 2002a. Development and application of siRNA expression vector. Nucleic Acids Res Suppl:113-114.

Miyagishi M, Taira K. 2002b. U6 promoter-driven siRNAs with four uridine 3' overhangs efficiently suppress targeted gene expression in mammalian cells. Nat Biotechnol 20:497-500.

Molnar A, Schwach F, Studholme DJ, Thuenemann EC, Baulcombe DC. 2007. miRNAs control gene expression in the single-cell alga Chlamydomonas reinhardtii. Nature 447:1126-1129.

Monaghan AP, Kaestner KH, Grau E, Schutz G. 1993. Postimplantation expression patterns indicate a role for the mouse forkhead/HNF-3 alpha, beta and gamma genes in determination of the definitive endoderm, chordamesoderm and neuroectoderm. Development 119:567-578.

Morini M, Piccini D, Barbieri O, Astigiano S. 1997. Modulation of alpha 6/beta 1 integrin expression during differentiation of F9 murine embryonal carcinoma cells to parietal endoderm. Exp Cell Res 232:304-312.

Moss EG, Lee RC, Ambros V. 1997. The cold shock domain protein LIN-28 controls developmental timing in C. elegans and is regulated by the lin-4 RNA. Cell 88:637-646.

Nagamine CM, Michot JL, Roberts C, Guenet JL, Bishop CE. 1987a. Linkage of the murine steroid sulfatase locus, Sts, to sex reversed, Sxr: a genetic and molecular analysis. Nucleic Acids Res 15:9227-9238.

Nagamine CM, Taketo T, Koo GC. 1987b. Studies on the genetics of tda-1 XY sex reversal in the mouse. Differentiation 33:223-231.

Nagy A, Gocza E, Diaz EM, Prideaux VR, Ivanyi E, Markkula M, Rossant J. 1990. Embryonic stem cells alone are able to support fetal development in the mouse. Development 110:815-821.

Nakagawa M, Koyanagi M, Tanabe K, Takahashi K, Ichisaka T, Aoi T, Okita K, Mochiduki Y, Takizawa N, Yamanaka S. 2008. Generation of induced pluripotent stem cells without Myc from mouse and human fibroblasts. Nat Biotechnol 26:101-106

Nichols J, Zevnik B, Anastassiadis K, Niwa H, Klewe-Nebenius D, Chambers I, Scholer H, Smith A. 1998. Formation of pluripotent stem cells in the mammalian embryo depends on the POU transcription factor Oct4. Cell 95:379-391.

Niederreither K, Dolle P. 2008. Retinoic acid in development: towards an integrated view. Nat Rev Genet 9:541-553.

Nikonova L, Koza RA, Mendoza T, Chao PM, Curley JP, Kozak LP. 2008. Mesoderm-specific transcript is associated with fat mass expansion in response to a positive energy balance. FASEB J.

Nishizaki Y, Shimazu K, Kondoh H, Sasaki H. 2001. Identification of essential sequence motifs in the node/notochord enhancer of Foxa2 (Hnf3beta) gene that are conserved across vertebrate species. Mech Dev 102:57-66.

Niwa H, Burdon T, Chambers I, Smith A. 1998. Self-renewal of pluripotent embryonic stem cells is mediated via activation of STAT3. Genes Dev 12:2048-2060.

Niwa H, Miyazaki J, Smith AG. 2000. Quantitative expression of Oct-3/4 defines differentiation, dedifferentiation or self-renewal of ES cells. Nat Genet 24:372-376.

Nussbaum J, Minami E, Laflamme MA, Virag JA, Ware CB, Masino A, Muskheli V, Pabon L, Reinecke H, Murry CE. 2007. Transplantation of undifferentiated murine embryonic stem cells in the heart: teratoma formation and immune response. FASEB J 21:1345-1357.

Okita K, Ichisaka T, Yamanaka S. 2007. Generation of germline-competent induced pluripotent stem cells. Nature 448:313-317.

Olsen PH, Ambros V. 1999. The lin-4 regulatory RNA controls developmental timing in Caenorhabditis elegans by blocking LIN-14 protein synthesis after the initiation of translation. Dev Biol 216:671-680.

Orban PC, Chui D, Marth JD. 1992. Tissue- and site-specific DNA recombination in transgenic mice. Proc Natl Acad Sci U S A 89:6861-6865.

Ota A, Tagawa H, Karnan S, Tsuzuki S, Karpas A, Kira S, Yoshida Y, Seto M. 2004. Identification and characterization of a novel gene C13orf25, as a target for 13q31-q32 amplification in malignant lymphoma. Cancer Res 64:3087-3095.

Otsu N. 1979. A threshold selection method from gray-level histograms. IEEE Trans. Sys., Man., Cyber 9:62–66.

Paddison PJ, Caudy AA, Bernstein E, Hannon GJ, Conklin DS. 2002. Short hairpin RNAs (shRNAs) induce sequence-specific silencing in mammalian cells. Genes Dev 16:948-958.

Papaioannou VE, McBurney MW, Gardner RL, Evans MJ. 1975. Fate of teratocarcinoma cells injected into early mouse embryos. Nature 258:70-73.

Park EJ, Sun X, Nichol P, Saijoh Y, Martin JF, Moon AM. 2008. System for tamoxifen-inducible expression of Cre-recombinase from the Foxa2 locus in mice. Dev Dyn 237:447-453.

Paul CP, Good PD, Winer I, Engelke DR. 2002. Effective expression of small interfering RNA in human cells. Nat Biotechnol 20:505-508.

Perea-Gomez A, Lawson KA, Rhinn M, Zakin L, Brulet P, Mazan S, Ang SL. 2001a. Otx2 is required for visceral endoderm movement and for the restriction of posterior signals in the epiblast of the mouse embryo. Development 128:753-765.

Perea-Gomez A, Rhinn M, Ang SL. 2001b. Role of the anterior visceral endoderm in restricting posterior signals in the mouse embryo. Int J Dev Biol 45:311-320.

Perea-Gomez A, Shawlot W, Sasaki H, Behringer RR, Ang S. 1999. HNF3beta and Lim1 interact in the visceral endoderm to regulate primitive streak formation and anterior-posterior polarity in the mouse embryo. Development 126:4499-4511.

Perea-Gomez A, Vella FD, Shawlot W, Oulad-Abdelghani M, Chazaud C, Meno C, Pfister V, Chen L, Robertson E, Hamada H, Behringer RR, Ang SL. 2002. Nodal antagonists in the anterior visceral endoderm prevent the formation of multiple primitive streaks. Dev Cell 3:745-756.

Piccolo S, Agius E, Leyns L, Bhattacharyya S, Grunz H, Bouwmeester T, De Robertis EM. 1999. The head inducer Cerberus is a multifunctional antagonist of Nodal, BMP and Wnt signals. Nature 397:707-710.

Piccolo S, Sasai Y, Lu B, De Robertis EM. 1996. Dorsoventral patterning in Xenopus: inhibition of ventral signals by direct binding of chordin to BMP-4. Cell 86:589-598.

Pillai RS, Bhattacharyya SN, Artus CG, Zoller T, Cougot N, Basyuk E, Bertrand E, Filipowicz W. 2005. Inhibition of translational initiation by Let-7 MicroRNA in human cells. Science 309:1573-1576.

Qi X, Li TG, Hao J, Hu J, Wang J, Simmons H, Miura S, Mishina Y, Zhao GQ. 2004. BMP4 supports self-renewal of embryonic stem cells by inhibiting mitogen-activated protein kinase pathways. Proc Natl Acad Sci U S A 101:6027-6032.

Raff M. 2003. Adult stem cell plasticity: fact or artifact? Annu Rev Cell Dev Biol 19:1-22.

Rajewsky K, Gu H, Kuhn R, Betz UA, Muller W, Roes J, Schwenk F. 1996. Conditional gene targeting. J Clin Invest 98:600-603.
Rand TA, Petersen S, Du F, Wang X. 2005. Argonaute2 cleaves the anti-guide strand of siRNA during RISC activation. Cell 123:621-629.
Rehwinkel J, Natalin P, Stark A, Brennecke J, Cohen SM, Izaurralde E. 2006. Genome-wide analysis of mRNAs regulated by Drosha and Argonaute proteins in *Drosophila melanogaster*. Mol Cell Biol 26:2965-2975.
Reinhart BJ, Slack FJ, Basson M, Pasquinelli AE, Bettinger JC, Rougvie AE, Horvitz HR, Ruvkun G. 2000. The 21-nucleotide *let-7* RNA regulates developmental timing in *Caenorhabditis elegans*. Nature 403:901-906.
Resnick JL, Bixler LS, Cheng L, Donovan PJ. 1992. Long-term proliferation of mouse primordial germ cells in culture. Nature 359:550-551.
Reya T, Morrison SJ, Clarke MF, Weissman IL. 2001. Stem cells, cancer, and cancer stem cells. Nature 414:105-111.
Rhinn M, Dierich A, Shawlot W, Behringer RR, Le Meur M, Ang SL. 1998. Sequential roles for Otx2 in visceral endoderm and neuroectoderm for forebrain and midbrain induction and specification. Development 125:845-856.
Robinson SR, Dreher B. 1990. The visual pathways of eutherian mammals and marsupials develop according to a common timetable. Brain Behav Evol 36:177-195.
Rodaway A, Patient R. 2001. Mesendoderm. an ancient germ layer? Cell 105:169-172.
Rodaway A, Takeda H, Koshida S, Broadbent J, Price B, Smith JC, Patient R, Holder N. 1999. Induction of the mesendoderm in the zebrafish germ ring by yolk cell-derived TGF-beta family signals and discrimination of mesoderm and endoderm by FGF. Development 126:3067-3078.
Rodriguez A, Vigorito E, Clare S, Warren MV, Couttet P, Soond DR, van Dongen S, Grocock RJ, Das PP, Miska EA, Vetrie D, Okkenhaug K, Enright AJ, Dougan G, Turner M, Bradley A. 2007. Requirement of *bic*/microRNA-155 for normal immune function. Science 316:608-611.
Rohwedel J, Maltsev V, Bober E, Arnold HH, Hescheler J, Wobus AM. 1994. Muscle cell differentiation of embryonic stem cells reflects myogenesis *in vivo*: developmentally regulated expression of myogenic determination genes and functional expression of ionic currents. Dev Biol 164:87-101.
Ronchetti D, Lionetti M, Mosca L, Agnelli L, Andronache A, Fabris S, Deliliers GL, Neri A. 2008. An integrative genomic approach reveals coordinated expression of intronic miR-335, miR-342, and miR-561 with deregulated host genes in multiple myeloma. BMC Med Genomics 1:37.
Ross SA, McCaffery PJ, Drager UC, De Luca LM. 2000. Retinoids in embryonal development. Physiol Rev 80:1021-1054.
Rossant J. 2001. Stem cells from the Mammalian blastocyst. Stem Cells 19:477-482.
Rossi SW, Jenkinson WE, Anderson G, Jenkinson EJ. 2006. Clonal analysis reveals a common progenitor for thymic cortical and medullary epithelium. Nature 441:988-991.
Sakamoto Y, Hara K, Kanai-Azuma M, Matsui T, Miura Y, Tsunekawa N, Kurohmaru M, Saijoh Y, Koopman P, Kanai Y. 2007. Redundant roles of *Sox17* and *Sox18* in early cardiovascular development of mouse embryos. Biochem Biophys Res Commun 360:539-544.
Sasaki H, Hogan BL. 1993. Differential expression of multiple *fork head* related genes during gastrulation and axial pattern formation in the mouse embryo. Development 118:47-59.
Sasaki H, Hogan BL. 1996. Enhancer analysis of the mouse *HNF-3 beta* gene: regulatory elements for node/notochord and floor plate are independent and consist of multiple sub-elements. Genes Cells 1:59-72.
Sasaki T, Shiohama A, Minoshima S, Shimizu N. 2003. Identification of eight members of the Argonaute family in the human genome small star, filled. Genomics 82:323-330.
Sauer B, Henderson N. 1988. Site-specific DNA recombination in mammalian cells by the Cre recombinase of bacteriophage P1. Proc Natl Acad Sci U S A 85:5166-5170.
Sayed D, Hong C, Chen IY, Lypowy J, Abdellatif M. 2007. MicroRNAs play an essential role in the development of cardiac hypertrophy. Circ Res 100:416-424.
Serls AE, Doherty S, Parvatiyar P, Wells JM, Deutsch GH. 2005. Different thresholds of fibroblast growth factors pattern the ventral foregut into liver and lung. Development 132:35-47.
Shamblott MJ, Axelman J, Wang S, Bugg EM, Littlefield JW, Donovan PJ, Blumenthal PD, Huggins GR, Gearhart JD. 1998. Derivation of pluripotent stem cells from cultured human primordial germ cells. Proc Natl Acad Sci U S A 95:13726-13731.
Shan G, Li Y, Zhang J, Li W, Szulwach KE, Duan R, Faghihi MA, Khalil AM, Lu L, Paroo Z, Chan AW, Shi Z, Liu Q, Wahlestedt C, He C, Jin P. 2008. A small molecule enhances RNA interference and promotes microRNA processing. Nat Biotechnol 26:933-940.
Shawlot W, Deng JM, Behringer RR. 1998. Expression of the mouse cerberus-related gene, *Cerr1*, suggests a role in anterior neural induction and somitogenesis. Proc Natl Acad Sci U S A 95:6198-6203.
Shen W, Scearce LM, Brestelli JE, Sund NJ, Kaestner KH. 2001. Foxa3 (hepatocyte nuclear factor 3gamma) is required for the regulation of hepatic GLUT2 expression and the maintenance of glucose homeostasis during a prolonged fast. J Biol Chem 276:42812-42817.

References

Shih J, Fraser SE. 1996. Characterizing the zebrafish organizer: microsurgical analysis at the early-shield stage. Development 122:1313-1322.

Shim EY, Woodcock C, Zaret KS. 1998. Nucleosome positioning by the winged helix transcription factor HNF3. Genes Dev 12:5-10.

Shimshek DR, Kim J, Hubner MR, Spergel DJ, Buchholz F, Casanova E, Stewart AF, Seeburg PH, Sprengel R. 2002. Codon-improved Cre recombinase (iCre) expression in the mouse. Genesis 32:19-26.

Shiraki T, Kondo S, Katayama S, Waki K, Kasukawa T, Kawaji H, Kodzius R, Watahiki A, Nakamura M, Arakawa T, Fukuda S, Sasaki D, Podhajska A, Harbers M, Kawai J, Carninci P, Hayashizaki Y. 2003. Cap analysis gene expression for high-throughput analysis of transcriptional starting point and identification of promoter usage. Proc Natl Acad Sci U S A 100:15776-15781.

Sinner D, Rankin S, Lee M, Zorn AM. 2004. Sox17 and beta-catenin cooperate to regulate the transcription of endodermal genes. Development 131:3069-3080.

Smith AG. 2001. Embryo-derived stem cells: of mice and men. Annu Rev Cell Dev Biol 17:435-462.

Smithies O, Gregg RG, Boggs SS, Koralewski MA, Kucherlapati RS. 1985. Insertion of DNA sequences into the human chromosomal beta-globin locus by homologous recombination. Nature 317:230-234.

Sokol NS, Ambros V. 2005. Mesodermally expressed Drosophila microRNA-1 is regulated by Twist and is required in muscles during larval growth. Genes Dev 19:2343-2354.

Sorensen DR, Leirdal M, Sioud M. 2003. Gene silencing by systemic delivery of synthetic siRNAs in adult mice. J Mol Biol 327:761-766.

Soriano P. 1999. Generalized lacZ expression with the ROSA26 Cre reporter strain. Nat Genet 21:70-71.

Spemann H, Mangold H. 1924. Über Induktion von Embryonalanlagen durch Implantation artfremder Organisatoren. Arch Mikr Anat Entwicklungsmech 100:599-638.

Srinivas S, Rodriguez T Clements M, Smith JC, Beddington RS. 2004. Active cell migration drives the unilateral movements of the anterior visceral endoderm. Development 131:1157-1164.

Stark A, Brennecke J, Bushati N, Russell RB, Cohen SM. 2005. Animal MicroRNAs confer robustness to gene expression and have a significant impact on 3'UTR evolution. Cell 123:1133-1146.

Stefani G, Slack FJ. 2008. Small non-coding RNAs in animal development. Nat Rev Mol Cell Biol 9:219-230.

Sternberg N, Hamilton D. 1981. Bacteriophage P1 site-specific recombination. I. Recombination between loxP sites. J Mol Biol 150:467-486.

Sternberg N, Hamilton D, Austin S, Yarmolinsky M, Hoess R. 1981. Site-specific recombination and its role in the life cycle of bacteriophage P1. Cold Spring Harb Symp Quant Biol 45 Pt 1 297-309.

Suh MR, Lee Y, Kim JY, Kim SK, Moon SH, Lee JY, Cha KY, Chung HM, Yoon HS, Moon SY, Kim VN, Kim KS. 2004. Human embryonic stem cells express a unique set of microRNAs. Dev Biol 270:488-498.

Sui G, Soohoo C, Affar el B, Gay F, Shi Y, Forrester WC, Shi Y. 2002. A DNA vector-based RNAi technology to suppress gene expression in mammalian cells. Proc Natl Acad Sci U S A 99:5515-5520.

Sund NJ, Vatamaniuk MZ, Casey M, Ang SL, Magnuson MA, Stoffers DA, Matschinsky FM, Kaestner KH. 2001. Tissue-specific deletion of Foxa2 in pancreatic beta cells results in hyperinsulinemic hypoglycemia. Genes Dev 15:1706-1715.

Tada S, Era T, Furusawa C, Sakurai H, Nishikawa S, Kinoshita M, Nakao K, Chiba T, Nishikawa S. 2005. Characterization of mesendoderm: a diverging point of the definitive endoderm and mesoderm in embryonic stem cell differentiation culture. Development 132:4363-4374.

Takada S, Stark KL, Shea MJ, Vassileva G, McMahon JA, McMahon AP. 1994. Wnt-3a regulates somite and tailbud formation in the mouse embryo. Genes Dev 8:174-189.

Takahashi K, Okita K, Nakagawa M, Yamanaka S. 2007a. Induction of pluripotent stem cells from fibroblast cultures. Nat Protoc 2:3081-3089.

Takahashi K, Tanabe K, Ohnuki M, Narita M, Ichisaka T, Tomoda K, Yamanaka S. 2007b. Induction of pluripotent stem cells from adult human fibroblasts by defined factors. Cell 131:861-872.

Takahashi K, Yamanaka S. 2006. Induction of pluripotent stem cells from mouse embryonic and adult fibroblast cultures by defined factors. Cell 126:663-676.

Takash W, Canizares J, Bonneaud N, Poulat F, Mattei MG, Jay P, Berta P. 2001. SOX7 transcription factor: sequence, chromosomal localisation, expression, transactivation and interference with Wnt signalling. Nucleic Acids Res 29:4274-4283.

Takashima S, Mkrtchyan M, Younossi-Hartenstein A, Merriam JR, Harterstein V. 2008. The behaviour of Drosophila adult hindgut stem cells is controlled by Wnt and Hh signalling. Nature 454:651-655.

Tam PP, Khoo PL, Lewis SL, Bildsoe H, Wong N, Tsang TE, Gad JM, Robb L. 2007. Sequential allocation and global pattern of movement of the definitive endoderm in the mouse embryo during gastrulation. Development 134:251-260.

Tam PP, Steiner KA, Zhou SX, Quinlan GA. 1997a. Lineage and functional analyses of the mouse organizer. Cold Spring Harb Symp Quant Biol 62:135-144.

Tam W, Ben-Yehuda D, Hayward WS. 1997b. Bic, a novel gene activated by proviral insertions in avian leukosis virus-induced lymphomas, is likely to function through its noncoding RNA. Mol Cell Biol 17:1490-1502.

Tam W, Hughes SH, Hayward WS, Besmer P. 2002. Avian bic, a gene isolated from a common retroviral site in avian leukosis virus-induced lymphomas that encodes a noncoding RNA, cooperates with c-myc in lymphomagenesis and erythroleukemogenesis. J Virol 76:4275-4286.

Tamplin OJ, Kinzel D, Cox B, Bell CE, Rossant J, Lickert H. 2008. Microarray analysis of *Foxa2* mutant mouse embryos identifies novel genes in the gastrula organizer and its derivates. BMC Genomics. In press.

Tavazoie SF, Alarcon C, Oskarsson T, Padua D, Wang Q, Bos PD, Gerald WL, Massague J. 2008. Endogenous human microRNAs that suppress breast cancer metastasis. Nature 451:147-152.

Thomas KR, Capecchi MR. 1986. Targeting of genes to specific sites in the mammalian genome. Cold Spring Harb Symp Quant Biol 51 Pt 2:1101-1113.

Thomas KR, Capecchi MR. 1987. Site-directed mutagenesis by gene targeting in mouse embryo-derived stem cells. Cell 51:503-512.

Thomas P, Beddington R. 1996. Anterior primitive endoderm may be responsible for patterning the anterior neural plate in the mouse embryo. Curr Biol 6:1487-1496.

Thomas PQ, Brown A, Beddington RS. 1998. *Hex*: a homeobox gene revealing peri-implantation asymmetry in the mouse embryo and an early transient marker of endothelial cell precursors. Development 125:85-94.

Thomson JA, Itskovitz-Eldor J, Shapiro SS, Waknitz MA, Swiergiel JJ, Marshall VS, Jones JM. 1998. Embryonic stem cell lines derived from human blastocysts. Science 282:1145-1147.

Thomson JA, Kalishman J, Golos TG, Durning M, Harris CP, Becker RA, Hearn JP. 1995. Isolation of a primate embryonic stem cell line. Proc Natl Acad Sci U S A 92:7844-7848.

Todaro GJ, Green H. 1963. Quantitative studies of the growth of mouse embryo cells in culture and their development into established lines. J Cell Biol 17:299-313.

Torihashi S. 2006. Formation of gut-like structures *in vitro* from mouse embryonic stem cells. Methods Mol Biol 330:279-285.

Torihashi S, Kuwahara M, Ogaeri T, Zhu P, Kurahashi M, Fujimoto T. 2006. Gut-like structures from mouse embryonic stem cells as an *in vitro* model for gut organogenesis preserving developmental potential after transplantation. Stem Cells 24:2618-2626.

Torres-Padilla ME. 2008. Cell identity in the preimplantation mammalian embryo: an epigenetic perspective from the mouse. Hum Reprod 23:1246-1252.

Tremblay KD, Hoodless PA, Bikoff EK, Robertson EJ. 2000. Formation of the definitive endoderm in mouse is a Smad2-dependent process. Development 127:3079-3090.

Uetzmann L, Burtscher I, Lickert H. 2008. A mouse line expressing *Foxa2*-driven Cre recombinase in node, notochord, floorplate, and endoderm. Genesis; 46(10):515-22.

van den Berg A, Kroesen BJ, Kooistra K, de Jong D, Briggs J, Blokzijl T, Jacobs S, Kluiver J, Diepstra A, Maggio E, Poppema S. 2003. High expression of B-cell receptor inducible gene *BIC* in all subtypes of Hodgkin lymphoma. Genes Chromosomes Cancer 37:20-28.

van Es JH, Jay P, Gregorieff A, van Gijn ME, Jonkheer S, Hatzis P, Thiele A, van den Born M, Begthel H, Brabletz T, Taketo MM, Clevers H. 2005. Wnt signalling induces maturation of Paneth cells in intestinal crypts. Nat Cell Biol 7:381-386.

Varlet I, Collignon J, Norris DP, Robertson EJ. 1997a. Nodal signaling and axis formation in the mouse. Cold Spring Harb Symp Quant Biol 62:105-113.

Varlet I, Collignon J, Robertson EJ. 1997b. *Nodal* expression in the primitive endoderm is required for specification of the anterior axis during mouse gastrulation. Development 124:1033-1044.

Varshavsky A. 1997. The N-end rule pathway of protein degradation. Genes Cells 2:13-28.

Vennstrom B, Sheiness D, Zabielski J, Bishop JM. 1982. Isolation and characterization of *c-myc*, a cellular homolog of the oncogene (*v-myc*) of avian myelocytomatosis virus strain 29. J Virol 42:773-779.

Verhave JC, Gansevoort RT, Hillege HL, De Zeeuw D, Curhan GC, De Jong PE. 2004. Drawbacks of the use of indirect estimates of renal function to evaluate the effect of risk factors on renal function. J Am Soc Nephrol 15:1316-1322.

Vincent SD, Dunn NR, Hayashi S, Norris DP, Robertson EJ. 2003. Cell fate decisions within the mouse organizer are governed by graded Nodal signals. Genes Dev 17:1646-1662.

Waddington CH, Waterman AJ. 1933. The Development *in vitro* of Young Rabbit Embryos. J Anat 67:355-370.

Wakitani S, Takaoka K, Hattori T, Miyazawa N, Iwanaga T, Takeda S, Watanabe TK, Tanigami A. 2003. Embryonic stem cells injected into the mouse knee joint form teratomas and subsequently destroy the joint. Rheumatology (Oxford) 42:162-165.

Wan H, Dingle S, Xu Y, Besnard V, Kaestner KH, Ang SL, Wert S, Stahlman MT, Whitsett JA. 2005. Compensatory roles of *Foxa1* and *Foxa2* during lung morphogenesis. J Biol Chem 280:13809-13816.

Wan H, Xu Y, Ikegami M, Stahlman MT, Kaestner KH, Ang SL, Whitsett JA. 2004. Foxa2 is required for transition to air breathing at birth. Proc Natl Acad Sci U S A 101:14449-14454.

Wang H, Dey SK. 2006. Roadmap to embryo implantation: clues from mouse models. Nat Rev Genet 7:185-199.

Warming S, Costantino N, Court DL, Jenkins NA, Copeland NG. 2005. Simple and highly efficient BAC recombineering using galK selection. Nucleic Acids Res 33:e36.

Waterston RH, Lindblad-Toh K, Birney E, Rogers J, Abril JF, Agarwal P, Agarwala R, Ainscough R, Alexandersson M, An P, Antonarakis SE, Attwood J, Baertsch R, Bailey J, Barlow K, Beck S, Berry E, Birren B, Bloom T, Bork P, Botcherby M, Bray N, Brent MR, Brown DG, Brown SD, Bult C, Burton J, Butler J, Campbell RD, Carninci P, Cawley S, Chiaromonte F, Chinwalla AT, Church DM, Clamp M, Clee C, Collins FS, Cook LL, Copley RR, Coulson A, Couronne O, Cuff J, Curwen V, Cutts T, Daly M, David R, Davies J, Delehaunty KD, Deri J, Dermitzakis ET, Dewey C, Dickens NJ, Diekhans M, Dodge S, Dubchak I, Dunn DM, Eddy SR, Elnitski L, Emes RD, Eswara P, Eyras E, Felsenfeld A, Fewell GA, Flicek P, Foley K, Frankel WN, Fulton LA, Fulton RS, Furey TS, Gage D, Gibbs RA, Glusman G, Gnerre S, Goldman N, Goodstadt L, Grafham D, Graves TA, Green ED, Gregory S, Guigo R, Guyer M, Hardison RC, Haussler D, Hayashizaki Y, Hillier LW, Hinrichs A, Hlavina W, Holzer T, Hsu F, Hua A, Hubbard T, Hunt A, Jackson I, Jaffe DB, Johnson LS, Jones M, Jones TA, Joy A, Kamal M, Karlsson EK, Karolchik D, Kasprzyk A, Kawai J, Keibler E, Kells C, Kent WJ, Kirby A, Kolbe DL, Korf I, Kucherlapati RS, Kulbokas EJ, Kulp D, Landers T, Leger JP, Leonard S, Letunic I, Levine R, Li J, Li M, Lloyd C, Lucas S, Ma B, Maglott DR, Marcis ER, Matthews L, Mauceli E, Mayer JH, McCarthy M, McCombie WR, McLaren S, McLay K, McPherson JD, Meldrim J, Meredith B, Mesirov JP, Miller W, Miner TL, Mongin E, Montgomery KT, Morgan M, Mott R, Mullikin JC, Muzny DM, Nash WE, Nelson JO, Nhan MN, Nicol R, Ning Z, Nusbaum C, O'Connor MJ, Okazaki Y, Oliver K, Overton-Larty E, Pachter L, Parra G, Pepin KH, Peterson J, Pevzner P, Flumb R, Pohl CS, Poliakov A, Ponce TC, Ponting CP, Potter S, Quail M, Reymond A, Roe BA, Roskin KM, Rubin EM, Rust AG, Santos R, Sapojnikov V, Schultz B, Schultz J, Schwartz MS, Schwartz S, Scott C, Seaman S, Searle S, Sharpe T, Sheridan A, Shownkeen R, Sims S, Singer JB, Slater G, Smit A, Smith DR, Spencer B, Stabenau A, Stange-Thomann N, Sugnet C, Suyama M, Tesler G, Thompson J, Torrents D, Trevaskis E, Tromp J, Ucla C, Ureta-Vidal A, Vinson JP, von Niederhausern AC, Wade CM, Wall M, Weber RJ, Weiss RB, Wendl MC, West AP, Wetterstrand K, Wheeler R, Whelan S, Wierzbowski J, Willey D, Williams S, Wilson RK, Winter E, Worley KC, Wyman D, Yang S, Yang SP, Zdobnov EM, Zody MC, Lander ES. 2002. Initial sequencing and comparative analysis of the mouse genome. Nature 420:520-562.

Wederell ED, Bilenky M, Cullum R, Thiessen N, Dagpinar M, Delaney A, Varhol R, Zhao Y, Zeng T, Bernier B, Ingham M, Hirst M, Robertson G, Marra MA, Jones HS, Hoodless PA. 2008. Global analysis of in vivo Foxa2-binding sites in mouse adult liver using massively parallel sequencing. Nucleic Acids Res 36:4549-4564.

Weigel D, Jackle H. 1990. The fork head domain: a novel DNA binding motif of eukaryotic transcription factors? Cell 63:455-456.

Weigel D, Jürgens G, Kuttner F, Seifert E, Jäckle H. 1989. The homeotic gene *fork head* encodes a nuclear protein and is expressed in the terminal regions of the Drosophila embryo. Cell 57:645-658.

Weinstein DC, Ruiz i Altaba A, Chen WS, Hoodless P, Prezioso VR, Jessell TM, Darnell JE, Jr. 1994. The winged-helix transcription factor HNF-3 beta is required for notochord development in the mouse embryo. Cell 78:575-588.

Wells JM, Melton DA. 1999. Vertebrate endoderm development. Annu Rev Cell Dev Biol 15:393-410.

Wells JM, Melton DA. 2000. Early mouse endoderm is patterned by soluble factors from adjacent germ layers. Development 127:1563-1572.

Wetzel R. 1925. Untersuchungen am Hühnerkeim. 1. Über die Untersuchungen des lebenden Keims mit neueren Methoden, besonders der Vogtschen vitalen Farbmarkierung. Wilhelm Roux Arch Entwicklungsmech 106:463–468.

Wienholds E, Koudijs MJ, van Eeden FJ, Cuppen E, Plasterk RH. 2003. The microRNA-producing enzyme Dicer1 is essential for zebrafish development. Nat Genet 35:217-218.

Wightman B, Ha I, Ruvkun G. 1993. Posttranscriptional regulation of the heterochronic gene *lin-14* by *lin-4* mediates temporal pattern formation in C. elegans. Cell 75:855-862.

Wiles MV, Vauti F, Otte J, Fuchtbauer EM, Ruiz P, Fuchtbauer A, Arnold HH, Lehrach H, Metz T, von Melchner H, Wurst W. 2000. Establishment of a gene-trap sequence tag library to generate mutant mice from embryonic stem cells. Nat Genet 24:13-14.

Wilkinson DG, Bhatt S, Herrmann BG. 1990. Expression pattern of the mouse T gene and its role in mesoderm formation. Nature 343:657-659.

Wilcoxon F. 1945. Individual comparisons by ranking methods. Biometrics 1:80-83.

Wodarz A, Nusse R. 1998. Mechanisms of Wnt signaling in development. Annu Rev Cell Dev Biol 14:59-88.

Wolfrum C, Asilmaz E, Luca E, Friedman JM, Stoffel M. 2004. Foxa2 regulates lipic metabolism and ketogenesis in the liver during fasting and in diabetes. Nature 432:1027-1032.

Wolfrum C, Besser D, Luca E, Stoffe M. 2003a. Insulin regulates the activity of forkhead transcription factor Hnf-3beta/Foxa-2 by Akt-mediated phosphorylation and nuclear/cytosolic localization. Proc Natl Acad Sci U S A 100:11624-11629.

Wolfrum C, Howell JJ, Ndungo E, Stoffel M. 2008. Foxa2 activity increases plasma high density lipoprotein levels by regulating apolipoprotein M. J Biol Chem 283:16940-16949.

Wolfrum C, Shih DQ, Kuwajima S, Norris AW, Kahn CR, Stoffel M. 2003b. Role of Foxa-2 in adipocyte metabolism and differentiation. J Clin Invest 112:345-356.

References

Wolfrum C, Stoffel M. 2006. Coactivation of *Foxa2* through Pgc-1beta promotes liver fatty acid oxidation and triglyceride/VLDL secretion. Cell Metab 3:99-110.

Wu KL, Gannon M, Peshavaria M, Offield MF, Henderson E, Ray M, Marks A, Gamer LW, Wright CV, Stein R. 1997. Hepatocyte nuclear factor 3beta is involved in pancreatic beta-cell-specific transcription of the *pdx-1* gene. Mol Cell Biol 17:6002-6013.

Wu L, Fan J, Belasco JG. 2006. MicroRNAs direct rapid deadenylation of mRNA. Proc Natl Acad Sci U S A 103:4034-4039.

Wurst W, Rossant J, Prideaux V, Kownacka M, Joyner A, Hill DP, Guillemot F, Gasca S, Cado D, Auerbach A, et al. 1995. A large-scale gene-trap screen for insertional mutations in developmentally regulated genes in mice. Genetics 139:889-899.

Yamada T, Yoshikawa M, Takaki M, Torihashi S, Kato Y, Nakajima Y, Ishizaka S, Tsunoda Y. 2002. *In vitro* functional gut-like organ formation from mouse embryonic stem cells. Stem Cells 20:41-49.

Yanaihara N, Caplen N, Bowman E, Seike M, Kumamoto K, Yi M, Stephens RM, Okamoto A, Yokota J, Tanaka T, Calin GA, Liu CG, Croce CM, Harris CC. 2006. Unique microRNA molecular profiles in lung cancer diagnosis and prognosis. Cancer Cell 9:189-198.

Yang B, Lin H, Xiao J, Lu Y, Luo X, Li B, Zhang Y, Xu C, Bai Y, Wang H, Chen G, Wang Z. 2007. The muscle-specific microRNA miR-1 regulates cardiac arrhythmogenic potential by targeting GJA1 and KCNJ2. Nat Med 13:486-491.

Yasunaga M, Tada S, Torikai-Nishikawa S, Nakano Y, Okada M, Jakt LM, Nishikawa S, Chiba T, Era T, Nishikawa S. 2005. Induction and monitoring of definitive and visceral endoderm differentiation of mouse ES cells. Nat Biotechnol 23:1542-1550.

Yi R, Doehle BP, Qin Y, Macara IG, Cullen BR. 2005. Overexpression of exportin 5 enhances RNA interference mediated by short hairpin RNAs and microRNAs. RNA 11:220-226.

Yi R, Qin Y, Macara IG, Cullen BR. 2003. Exportin-5 mediates the nuclear export of pre-microRNAs and short hairpin RNAs. Genes Dev 17:3011-3016.

Yu J, Vodyanik MA, Smuga-Otto K, Antosiewicz-Bourget J, Frane JL, Tian S, Nie J, Jonsdottir GA, Ruotti V, Stewart R, Slukvin, II, Thomson JA. 2007. Induced pluripotent stem cell lines derived from human somatic cells. Science 318:1917-1920.

Yu JY, DeRuiter SL, Turner DL. 2002. RNA interference by expression of short-interfering RNAs and hairpin RNAs in mammalian cells. Proc Natl Acad Sci U S A 99:6047-6052.

Yu JY, Taylor J, DeRuiter SL, Vojtek AB, Turner DL. 2003. Simultaneous inhibition of GSK3alpha and GSK3beta using hairpin siRNA expression vectors. Mol Ther 7:228-236.

Zaret KS. 2008. Genetic programming of liver and pancreas progenitors: lessons for stem-cell differentiation. Nat Rev Genet 9:329-340.

Zeng Y, Cullen BR. 2003. Sequence requirements for micro RNA processing and function in human cells. RNA 9:112-123.

Zeng Y, Cullen BR. 2004. Structural requirements for pre-microRNA binding and nuclear export by Exportin 5. Nucleic Acids Res 32:4776-4785.

Zeng Y, Cullen BR. 2005. Efficient processing of primary microRNA hairpins by Drosha requires flanking nonstructured RNA sequences. J Biol Chem 280:27595-27603.

Zeng Y, Yi R, Cullen BR. 2003. MicroRNAs and small interfering RNAs can inhibit mRNA expression by similar mechanisms. Proc Natl Acad Sci U S A 100:9779-9784.

Zeng Y, Yi R, Cullen BR. 2005. Recognition and cleavage of primary microRNA precursors by the nuclear processing enzyme Drosha. EMBO J 24:138-148.

Zhang L, Rubins NE, Ahima RS, Greenbaum LE, Kaestner KH. 2005. Foxa2 integrates the transcriptional response of the hepatocyte to fasting. Cell Metab 2:141-148.

Zhao T, Li G, Mi S, Li S, Hannon GJ, Wang XJ, Qi Y. 2007a. A complex system of small RNAs in the unicellular green alga *Chlamydomonas reinhardtii*. Genes Dev 21:1190-1203.

Zhao Y, Ransom JF, Li A, Vedantham V, von Drehle M, Muth AN, Tsuchihashi T, McManus MT, Schwartz RJ, Srivastava D. 2007b. Dysregulation of cardiogenesis, cardiac conduction, and cell cycle in mice lacking miRNA-1-2. Cell 129:303-317.

Zhao Y, Samal E, Srivastava D. 2005. Serum response factor regulates a muscle-specific microRNA that targets *Hand2* during cardiogenesis. Nature 436:214-220.

Zhou J, Ou-Yang Q, Li J, Zhou XY, Lin G, Lu GX. 2008. Human feeder cells support establishment and definitive endoderm differentiation of human embryonic stem cells. Stem Cells Dev 17:737-749.

Zhou X, Sasaki H, Lowe L, Hogan BL, Kuehn MR. 1993. *Nodal* is a novel TGF-beta-like gene expressed in the mouse node during gastrulation. Nature 361:543-547.

Zuk PA, Zhu M, Ashjian P, De Ugarte DA, Huang JI, Mizuno H, Alfonso ZC, Fraser JK, Benhaim P, Hedrick MH. 2002. Human adipose tissue is a source of multipotent stem cells. Mol Biol Cell 13:4279-4295.

8. Appendix

8.1. Publications

Uetzmann L, Burtscher I, Lickert H 2008. A mouse line expressing *Foxa2*-driven Cre recombinase in node, notochord, floorplate, and endoderm. Genesis. 2008 Oct;46(10):515-22.

Liao P, Uetzmann L, Burtscher I, Lickert H. under revision. Generation of a Mouse Line Expressing *Sox17*-driven Cre Recombinase with Specific Activity in Arteries. Genesis. 2009 Jul 47(7):476-83.

8.2. List of abbreviations

A	ADE	anterior definitive endoderm
	AIP	anterior intestinal port
	APC	activated protein c
	APS	6-amino-9-[3,4-dihydroxy-5-[(hydroxy-sulfooxy-phosphoryl) oxymethyl]oxolan-2-yl]purine
	ATP	adenosine triphosphate
	AVE	anterior visceral endoderm
B	BAC	bacterial artificial chromosome
	BCA	bicinchoninic acid
	BF	brightfield
	BMP	*bone morphogenic protein*
C	CAGE	cap analysis gene expression
	cDNA	copy DNA
	Cdx2	*caudal type homeobox 2*
	Cerl	*Cerberus like*
	CFP	cyan fluorescent protein
	c-Myc	*myelocytomatosis oncogene*
	CIP	caudal intestinal port
	Cre	causes recombination
	Cripto	*teratocarcinoma-derived growth factor 1 (Tdgf1)*
	CTP	cytosine triphosphate
D	DE	definitive endoderm
	dist.	distilled
	Dkk1	*Dickkopf 1*
	D. melanogaster	*Drosophila melanogaster*
	DMEM	Dulbecco's modified eagles medium
	DNA	Desoxyribonucleic acid
	dNTP(s)	desoxyriboucleotides
	Dsh	*Dishevelled*
	DVE	distal visceral endoderm
E	E	endoderm
	E. coli	*Escherichia coli*
	EB	embryonic body
	EDTA	ethylenediaminetetraacetic acid
	EG	embryonic germ (cells)
	EGO	early gastrula organizer
	EGTA	ethylene glycol tetraacetic acid
	ER	estrogen receptor
	ERT2	estrogen receptor type 2
	ES	embryonic stem
F	*Fgf*	*fibroblast growth factor*
	Flp	Flippase

F	Flp-e	Flp-enhanced
	Foxa2	*Forkheadbox transcription factor a2*
	FRT	*Flp-recombinase target*
	Frz	*Frizzled*
	fwd	forward
G	g	gravitation
	G418	Geneticin
	Gata4	*GATA binding protein 4*
	GFP	green fluorescent protein
	Gsc	*Goosecoid*
	GSK-3	glycogen synthase kinase
	GTP	gyanine triphosphate
H	H2B	Histon 2B
	HCl	hydrochlorid acid
	HEK293T	human embryonic kidney cell line (see Material and Methods)
	HEPES	N-2-Hydroxyethylpiperazine-N'-2-ethanesulfonic acid
	Hesx1	*homeo box gene expressed in ES cells*
	Hex	*Hematpoietically expressed homeobox gene*
	HMG	high mobility group
	Hox	*Homeobox*
	HPLC	high pressure liquid chromatography
	Hprt	Hypoxanthine-guanine phosphoribosyltransferase
	HR	homology region
	HRP	horse raddish peroxidase
I	ICM	inner cell mass
	iCre	*improved Cre*
	IFAB-P	*intestinal fatty acid binding protein*
	iPS	induced pluripotent stem cells
	IRES	internal ribosomal entry site
	IVD	in vitro differentiation
K	*K19*	*Cytokeratin 19*
	Kif4	*kinesin family member 4*
	Kozak	*Kozak sequence*
L	LB-medium	Luria Bertani medium
	LGO	late gastrula organizer
	LEF	*lymphoid enhancer binding factor*
	Lefty	*left right determination factor*
	LIF	leukemia inhibitory factor
	Lim1	*LIM homeobox protein 1 (Lhx1)*
	Lin28	*lin-28 homolog*
	loxP	*locus of X over Pi*
M	*MAP*	*mitogen-activated protein*
	MEF	murine embryonic fibroblasts
	MGO	mid-gastrula organizer
	miR	miRNA
	miRNA	micro RNA
	mRNA	messenger RNA
N	*Nanog*	*Nanog homeobox*
	neo	neomycin resistence
	NFP	(tumor) necrosis factor protection
O	*Oct4*	*Octamer-binding protein 4*
	ON	over night
	OH-group	alcohol group

O	ORF	open reading frame
	Otx1/2	*orthodenticle homolog 1/2*
P	P	Paternal
	P1	buffer P1 for plasmid preparation
	P2	buffer P2 for plasmid preparation
	P3	buffer P3 for plasmid preparation
	pA	polyadenylation signal
	pBKS-	Bluescript bacterial vector
	PBS	phosphate buffered saline
	PCR	polymerase chain reaction
	PGK	phosphoglycerate kinase
	PI	protease inhibitor
	Puro	puromycin resistance
R	RA	retinoic acid
	rev	reverse
	RNA	ribonucleic acid
	RNAi	RNA interference
	R26R	lacZ-reporter gene expressed from the ROSA locus, mouse line
	RT	room temperature
	RT-PCR	reverse transcription PCR
	rpm	rounds per minute
S	SIBR	synthetic inhibitory BIC-derived RNA
	siRNA	small-interfering RNA
	Sox17	*Sry-related HMG-box transcription factor 17*
	STAT	signal transducer and activator of transcription
	Sry	*sex determining region y*
	SV40	simian virus 40
T	TCF	T-cell factor
	Tgf	*transforming growth factor*
	T::GFP	GFP expressed from the T(Brachyury) locus
	T4 PNK	T4 virus polynucleotide kinase
	TTP	tyrosine triphosphate
U	UTR	untranslated region
V	VE	visceral endoderm
W	*Wnt*	*Wingless/Int*
X	X-Gal	5-bromo-4-chloro-3-indolyl β-D-galactoside

8.3. Listing of figures

Introduction

Figure 1: The first steps in embryonic development – reaching the blastula stage
Figure 2: Embryonic development of the mouse until implantation
Figure 3: Gastrulation in the mouse
Figure 4: Turning of the mouse embryo
Figure 5: The three principle germ layers and their derivatives
Figure 6: Anterior-posterior positioning of the endodermal cells and the link to their fate according to Lawson *et al.*, 1986 and 1991 (modified)
Figure 7: The embryonic gut – budding of the organs
Figure 8: Canonical Wnt signalling

Appendix

Figure 9: The genomic locus of *Foxa2*
Figure 10: The genomic locus of *Sox17*
Figure 11: The miRNA pathway

Results
Figure 12: Lineage tracing – identifying the progeny of progenitor cells – and studying the impact of signalling pathways – conditional knock-out analysis
Figure 13: Targeting strategy of the *Sox17^{iCre}* allele
Figure 14: Southern Blot on Sox17 targeted ES cells and mice - Verification of homologously recombined clones
Figure 15: Recombination activity of *Sox17^{iCre}* in the *ROSA* locus at E9.5
Figure 16: β-galactosidase activity in organs of *Sox17$^{iCre/+}$*; *R26$^{R/+}$* mice at P1
Figure 17: Targeting strategy of the *Foxa2^{iCre}* allele
Figure 18: Southern Blot on *Foxa2* targeted ES cells and mice - Verification of homologously recombined clones
Figure 19: Foxa2-iCre recombination activity at embryonic stages
Figure 20: Recombination activity of Foxa2-iCre in embryonic organs
Figure 21: Recombination activity of Foxa2-iCre in the embryonic heart
Figure 22: Histological sections for β-galactosidase activity of organs of *Foxa2$^{iCre/+}$*; *R26$^{R/+}$* mice
Figure 23: Analysis of Foxa2 protein levels in wild type versus *Foxa2$^{iCre/+}$* and *Foxa2$^{iCre/iCre}$* liver extracts at P1 using Western blot
Figure 24: Genomatix promoter prediction for *Foxa2*
Figure 25: Primers for RT-PCR on RNA samples from different tissues of *Foxa2$^{iCreΔneo/iCreΔneo}$*, *Foxa2$^{iCre/+}$* and *Foxa2$^{+/+}$* mice
Figure 26: Analysis of RNA samples from lung and liver of *Foxa2$^{iCreΔneo/iCreΔneo}$*, *Foxa2$^{iCreΔneo/+}$* and *Foxa2$^{+/+}$* mice regarding different possible transcripts using RT-PCR
Figure 27: Analysis of the Foxa2 protein concentration in samples from lung and pancreas of *Foxa2$^{iCreΔneo/iCreΔneo}$*, *Foxa2$^{iCreΔneo/+}$* and *Foxa2$^{+/+}$* mice using Western blot
Figure 28: Crossing scheme conditional *β-catenin* knock out A
Figure 29: Crossing scheme conditional *β-catenin* knock out B
Figure 30: E10.5 *Foxa2$^{iCre/+}$*; *β-catenin$^{floxdel/flox}$*; *R26$^{R/+}$* mice
Figure 31: Organs of *Foxa2$^{iCre/+}$*; *β-catenin$^{floxdel/flox}$*; *R26$^{R/+}$* mice at P5
Figure 32: Protocol for differentiation of ES cells into blood sugar dependent insulin-producing cells
Figure 33: ES differentiation according to Tada *et al.* (2005)
Figure 34: Optimization of differentiation using a co-culture system with NIH3T3 cells over-expressing Wnt3a
Figure 35: RT-PCR on RNA samples of *in vitro* differentiated ES cells
Figure 36: Live imaging of the onset of marker gene expression
Figure 37: Fluorescent fusion proteins as a read out of miRNA activity
Figure 38: Alteration of the expression vector for miRNAs and miRNA design
Figure 39: List of miRNAs used including folding predictions
Figure 40: miRNA transgenic ES cell clone (example)
Figure 41: *In vitro* differentiation of miRNA transgenic ES cell clones
Figure 42a: Foxa2 immunostaining on differentiated miRNA transgenic ES cell clones at d6 of differentiation
Figure 42b: Foxa2 immunostaining on differentiated miRNA transgenic ES cell clones at d6 of differentiation
Figure 43: Comparison of miR335, miR335 *Sox17*-optimized and miR194
Figure 44: Expanded model of endoderm differentiation *in vitro*: A model of the suppression and support of the establishment of endoderm versus mesoderm by miR335

Appendix

Discussion
Figure 45: Villus cell replacement in the adult small intestine
Figure 46: Model of a feedback loop of Foxa2 and miR335

Material and Methods
Figure 47: Southern blot setup

8.4. Listing of tables

Results
Table 1: Blood analysis of $Foxa2^{iCre\Delta neo/iCre\Delta neo}$, $Foxa2^{iCre\Delta neo/+}$ and $Foxa2^{+/+}$ mice
Table 2: Plasma analysis of $Foxa2^{iCre\Delta neo/iCre\Delta neo}$, $Foxa2^{iCre\Delta neo/+}$ and $Foxa2^{+/+}$ mice
Table 3: Blood analysis of $Foxa2^{iCre\Delta neo/+}$; $\beta\text{-}catenin^{flox/flox}$ mice
Table 4: Plasma analysis of $Foxa2^{iCre\Delta neo/+}$; $\beta\text{-}catenin^{flox/flox}$ mice

Material and Methods
Table 5: Oligonucleotides

8.5. Listing of charts

Results
Chart 1: Results of Mendelian distribution of $Foxa2^{iCre/iCre}$ mice
Chart 2: Preliminary data of Mendelian distribution of $Foxa2^{iCre/+}$; $\beta\text{-}catenin^{flox/flox}$ mice
Chart 3: Mendelian distribution of $Foxa2^{iCre/+}$; $\beta\text{-}catenin^{floxdel/flox}$; $R26^{R/+}$ mice
Chart 4: FACS analysis of the onset of marker gene expression

Die VDM Verlagsservicegesellschaft sucht für wissenschaftliche Verlage abgeschlossene und herausragende

Dissertationen, Habilitationen, Diplomarbeiten, Master Theses, Magisterarbeiten usw.

für die kostenlose Publikation als Fachbuch.

Sie verfügen über eine Arbeit, die hohen inhaltlichen und formalen Ansprüchen genügt, und haben Interesse an einer honorarvergüteten Publikation?

Dann senden Sie bitte erste Informationen über sich und Ihre Arbeit per Email an *info@vdm-vsg.de*.

Sie erhalten kurzfristig unser Feedback!

VDM Verlagsservicegesellschaft mbH
Dudweiler Landstr. 99
D - 66123 Saarbrücken

Telefon +49 681 3720 174
Fax   +49 681 3720 1749

www.vdm-vsg.de

Die VDM Verlagsservicegesellschaft mbH vertritt

Printed by Books on Demand GmbH, Norderstedt / Germany